Virtue and Responsibility in Policy Research and Advice

Berry Tholen

Virtue and Responsibility in Policy Research and Advice

palgrave
macmillan

Berry Tholen
Institute for Management Research,
Department of Public
Administration
Radboud University Nijmegen
Nijmegen, The Netherlands

ISBN 978-3-319-65252-8 ISBN 978-3-319-65253-5 (eBook)
DOI 10.1007/978-3-319-65253-5

Library of Congress Control Number: 2017949213

Printed on acid-free paper

This Palgrave Macmillan imprint is published by Springer Nature
The registered company is Springer International Publishing AG
The registered company address is: Gewerbestrasse 11, 6330 Cham, Switzerland

Preface

In recent years, values and virtues in scientific practice appear to have been rediscovered. Within and outside academia, concern about scientific fraud, data manipulation, plagiarism and the like has increased enormously. In all disciplines, in all faculties, and in all institutes for academic research, new measures have been initiated—for example, workshops on research ethics, hot lines to report abuses, special courses, codes of conduct, more transparency, and stricter systems of peer review. Some blame the typical career system, which demands publications and visibility, for the (alleged) increase in fraud, data trimming, and the like. Others refer to the perverting effects of the need to attract external funds to do research and the quick results that commissioning parties demand. All appear to agree, however, that some *ethical reveil* in science is in place; the relevance of values and virtues must be reconfirmed and elaborated. This book agrees with this latter conclusion, but for a completely different reason.

The concern here is not about doing methodologically sound research or about what might possibly threaten that activity. The subject of this book is the relationship between science and politics, between scientist advisors and policymakers. According to many observers, this relationship has been changing lately. Some identify a scientification of policymaking, that is, an increase in technocratic thinking on political issues. Others note a politicization of science by which they mean that political or other partisan interests ever more determine scientific practice and outcomes (e.g., Jasanoff 1990; Letsch & Weingart 2011). Both types of commentators agree, however, that there is a problem of one side intruding upon the sphere of the other. These critics of contemporary developments maintain that each sphere should be able to function autonomously, following its own type of rationality. Policymaking belongs to the sphere of value-deliberation; it should not be forced into the mould of technical reasoning. Science, conversely, should be free from values and politically neutral.

In this book, the ideal of a value-free science will be contested. It is argued that to be a good scientist, more is needed than avoiding fraud and data manipulation and knowing how to create valid knowledge. Good scientists are also involved in making social and political value-judgments. They must in fact possess a certain virtue at doing so.

Within the fields of Public Administration and Policy Analysis, the concern for values in research and in scientific advice is not new. The issue has been addressed, in one way or another, as part of a variety of subjects. It arises in discussions about the type of knowledge these disciplines can provide (e.g., in the Waldo-Simon debate). It is often active in the background of debates on interpretive versus positivist research. It was at issue in particular comments of mainstream Public Administrative research (as in the Minnowbrook Manifesto). Many more examples can be given. Here, the ideal of a value-free science and scientific policy advice will be addressed head-on. To do so, findings from policy analysis, the philosophy of science, and ethics will be brought together. The formal position of scientific advisors, and therefore their dependency upon the political agent they advise, can take many different forms. The advisor might be a completely independent (university) researcher, a contract researcher, a partner in a think tank, or a government employee. The particular type of relationship might bring all types of particular problems and opportunities (see for instance Bijker et al. 2009; Vining & Weimer 2017; Letsch & Weingart 2011). In this book, these typical principal-agent issues will be neglected. It focusses instead on the general aspects and complications of the relationship between scientific advisor and policymaker.

In addressing these issues, the argument of this book connects to an older research-tradition that encompasses Plato, Machiavelli, Weber, and many others who have analysed the relationship between knowledge and power or between science and politics. Its intention is not to offer a chronological overview of classical authors; nor does it want to draw a complete picture of all the issues and positions in this field. It focusses on a particular problem and tries to address it systematically. Of course, that does not mean that Weber and the others do not turn up in the argument.

This book is not a report of an empirical investigation. There is no attempt to verify reported trends (such as those on an increase in scientific fraud or the politicization of science); nor is there any intention to explain such phenomena. In this book, arguments and positions are discussed; it questions presuppositions and investigates logical consistency. The focus is not on what actually occurs, on what scientists do or on how that can be explained but rather on the question of what good scientific policy advisors should do.

In its intention, this book, at least in one aspect, resembles critical analyses of dominant discourses. It tries to bring into the light aspects of scientific advice that are often neglected, things that currently appear to be overlooked in the self-understanding of many scientists. Contrary to critical approaches, however, this book does try to unveil hidden power-positions or processes. Furthermore, the argument in this book does not challenge the scientific ambition of creating valid knowledge about the empirical world. In fact, it endorses that ambition.

A word on scientific knowledge. In this book, no particular scientific epistemology is defended. It merely presupposes, without further argument, that knowledge of the observable world is possible. More precisely, it takes as given the possibility of distinguishing better from worse beliefs about how the observable world actually is. These beliefs might concern causal theories or interpretations of meaning. Arguments for a radical scepticism or relativism about the possibilities of science, or knowledge in general, are neglected in this book.

The issues that this book addresses are of relevance for scholars that are involved in policy advice, but also for students and policymakers. Members from these different audiences might be more accustomed to some elements of this book than with others. For readers who are already familiar with the arguments that are made in the first part: the second part of the book can be read independently.

I want to thank Ringo Ossewaarde, Patrick Overeem, Sandra van Thiel, and Pieter Zwaan for the discussions we had on issues that found their way into this book and for their comments on (parts of) this book. They have contributed to its strengths; its flaws remain my responsibility. Also the participants in the Netherlands Institute of Government Ph.D.-course on responsibility in research and advice that I have been teaching over the past three years must be mentioned here: their enthusiasm, curiosity, and critical questions have stimulated and helped me to hammer out the argument of this book. I also thank Oxford University Press for its kind permission to adapt, in Chap. 7 of this book, elements of my article "The Value of the Issue Context Approach for Scientific Policy Advice", which appeared in *Science and Public Policy* 43(2016)2: 101–114.

Nijmegen, The Netherlands Berry Tholen

References

Bijker, W., Bal, R., & Hendriks, R. (2009). *The paradox of scientific authority. The role of scientific advice in democracies.* Cambridge, MA: MIT Press.

Jasanoff, S. (1990). *The Fifth Branch: Science Advisers as Policymakers.* Cambridge, MA: Harvard University Press.

Letsch, J., & Weingart, P. (Eds.). (2011). *The politics of scientific advice. Institutional design for quality assurance.* Cambridge: Cambridge University Press.

Vining, A. R., & Weimer, D. (2017). Policy Analysis: A valuable skill for public administrators. In J. C. N. Raadschelders & R. J. Stillman II (Eds.), *Foundations of public administration* (pp. 162–176). Irvine, CA: Melvin & Leigh.

Contents

CHAPTER 1

The Ideal of Value Neutrality in Policy Research and Advice: Some Inconvenient Observations

Abstract For many scientific policy advisors, as for most critics of contemporary scientific advisory practice, value freedom or political neutrality is the guiding ideal. Max Weber is often presented as the champion of this position. On closer inspection, however, this championship of value freedom becomes puzzling. Weber's claims on intellectual honesty, on studying what is worth knowing and on presenting inconvenient truths to policymakers urge investigation of the ideal of value neutrality. What type of value decisions must an expert advisor make, and how can he best address those issues?

Keywords Max weber · Value freedom · Neutrality · Inconvenient truth

1 Introduction

Below are two random examples of scholars who react in disbelief and even feel offended when they are accused of taking a political or ethical side in their research. The fierceness of their reaction shows that they feel their integrity as scientists is being questioned. More precisely, their understanding of what scientific integrity means is questioned:

> A set of anthropologists and sociologists studying Muslim communities in western countries find themselves criticized as being 'Islam-friendly' and

B. Tholen, *Virtue and Responsibility in Policy Research and Advice*,
DOI 10.1007/978-3-319-65253-5_1

> even as 'useful idiots'. One critic claims, for instance, that these scholars neglect the anti-democratic discourse of Salafism in their studies of the meaning that religion has for Islamic people. One of the accused researchers reacts in astonishment: 'These critics want me to make normative claims.' She explains that it is the task of a scientist to set aside his own judgments. He must try to understand why people act as they do and what motivates their choices, not judge them. (*NRC Handelsblad* 2016b)

> A legal scholar has performed a study on the possibility of legalizing the production and sale of cannabis under international law. Some of his colleagues claim that his study, given its particular research question, is bound to support the political position of government. A different research question might have delivered results that are far more critical of the government's policy. The scholar, in defence, replies that it absolutely is not his intention to choose sides in the political dispute. He simply accepted the research assignment and presented a thorough study answering the given question according to the best scientific methods. (*NRC Handelsblad* 2015)

For the scholars in these examples, doing research does not involve making political value decisions. It is about applying the technical methods of your trade, whether they are the methodological instruments of legal analysis, interpretive study or theoretical explanation. Scientists can provide empirical knowledge that is useful for policymaking, but they themselves should refrain from taking any political position or making any ethical valuation.

The usefulness of this type of politically neutral science is forcefully advocated by critics of 'fact-free politics'. Thus, these commentators demean policymaking based on unchecked or shallow empirical claims. Policymaking could be much improved if politicians and other policymakers actually considered the factual empirical knowledge that scientists provide. As impartial outsiders, these expert advisors can offer input for policies that can effectively solve problems. When their knowledge remains neglected, there is a risk that policies become mere symbolic devices. Scholars should therefore ensure that their input is considered (*NRC Handelsblad* 2016a; *Die Zeit* 2016).

The leading ideal in these examples is the political neutrality of scientists as policy advisors. It is an ideal that most likely is broadly shared among scientists. However, this ideal is not unchallenged, as the examples show. Can scientists, and scientist advisors more particularly, really be politically and ethically neutral? If not, what does that mean for the position of the scientist as a critical but non-partisan outsider? To address these questions, we will initially turn to the work of Max Weber. Weber offers a classical argument for the ideal of an apolitical or value-free social science (Fischer 1980: 21).

2 Weber's Intellectually Honest Social Scientist and Advisor

Max Weber developed his advocacy for an ethically neutral or value-free social science in an ongoing debate in the early years of the twentieth century in the German Sociological Association. The central issue of this discussion was how, and to what extent, scientific findings could support societal ideals and particular social policies and institutions. Some of Weber's colleagues maintained that social science could and should support judgments on the desirability of particular social policies and institutions. Some of them explicitly advocated socialist, communist, or more right-wing politics, suggesting that such value positions could be given a scientific grounding. They opted for a *wertende Soziologie* (normative or ethical social science). Weber agreed with his colleagues that science could offer a useful contribution to policymaking. However, against the advocates of a *wertende Soziologie*, Weber defended an ethically neutral, value-free social science.

Scientists, Weber maintains, should not suggest that value-judgments—that is, practical evaluations of phenomena—can have a base in science. They should completely refrain from political or ethical position-taking in the classroom (Weber 1948a: 145; 1949a: 2). In public debate, they should be clear when they are giving their personal opinions and (not scientifically supported) value-judgments. As Weber himself states, "What is really at issue is the intrinsically simple demand that the investigator and teacher should keep unconditionally separate the establishment of empirical facts [...] and his own practical evaluations, i.e., his evaluation of these facts as satisfactory or unsatisfactory" (Weber 1949a: 5, 11).

The argument that Weber provides for abstaining from value-judgments in science is that scientists should not suggest doing what they cannot do. It is logically impossible to decide on purely scientific grounds what value, which ideal end-state, or what political ideal is of the highest value and should prevail over others. Empirical science can present to us the state of facts in the world (how the world is). Answering normative questions (how the world should be) is, however, a completely different affair. These two issues are heterogeneous problems that belong to different categories, and each one demands its own approach. Deducing evaluative statements from strictly empirical ones is logically impossible. It is a matter of the intellectual honesty and integrity of a scientist to respect this fact-value distinction. Although it might occasionally be difficult to uphold this contrast, it should always be a guide in lecturing, research, and in giving advice (Weber 1948a: 146; 1949a: 1, 2, 9).

Weber concedes that scientists value doing research; without such a motivation, hardly any scientific enterprise would be undertaken. The results of this work, moreover, have their significance for us. Every scientific discipline is underpinned by some notion of what is of value. The general supposition of the medical scientific enterprise, for instance, is that its main tasks are maintaining life and diminishing suffering. The historical and cultural sciences presuppose that it is worthwhile to know and understand social phenomena. However, given its logical limitations, science cannot justify why these things are 'worth knowing' and why investigating them should be its task (Weber 1948a: 142–145).

The above does not mean, however, that a scientist cannot study values. Weber emphasizes that values, and value-judgments for that matter, are social or cultural phenomena that can be described and investigated empirically. An example might be a study of the values that guide police officers in their daily work and explain the choices they make. Values and value-judgments, then, are understood as objects of empirical study, just like policy outcomes or the working procedures of bureaucrats. For the social scientist, values are phenomena that can be counted and measured or described and interpreted. Their existence in a particular time and place might be an independent factor for explaining other social happenings. Moreover, their form and occurrence at any particular time and place can be the subject of an explanatory theory or a value-interpretation (Weber 1949b: 143). However, scientists should refrain from making value judgments about the cultural phenomena they study (Weber 1949a: 1). They should remain ethically neutral.

Occasionally, the position a scientist takes might appear to express the ideal of ethical neutrality but in fact does not mirror that ideal. One of Weber's examples concerns a scientist who associates himself with the values of the powers that be or the dominant interest groups. A second example concerns the scientist who opts for a value-position somewhere between competing ethical or political positions (a 'middle ground'). In both cases, the ideal of ethical neutrality is not realized. (In the latter case, what makes taking the middle ground in any dispute the best or neutral position?). In fact, each case shows partisanship for which—logically—no scientific ground can be provided (Weber 1949a: 6, 10).

Real ethical neutrality in science does not mean, however, that the scientists' work cannot be of help for policymakers or that it cannot be of importance for students' value-formation. Scientists, as academic teachers and as policy advisors, can and should present inconvenient facts to their audience, Weber maintains. These facts are inconvenient for their policy or party opinions. Empirical scientific findings might imply valued aims or desired end-states as infeasible or imply that they can only be realized by certain aims at high costs and with undesired side effects. Confronted with these inconvenient truths, policymakers might be persuaded to re-evaluate not only the means of their choice but also their aims. Knowledge of feasibility and consequences, then, can bring about a reorientation of priorities of values. "Each new fact may necessitate the re-adjustment of the relationships between ends and indispensable means, between desired goals and unavoidable subsidiary consequences" (Weber 1948a:151; 1949a: 23, 24). For example, policymakers, advised that the legal instruments they wished to employ to reduce the trade in soft drugs would not be effective, might decide to focus instead on trying to realize a decrease in the use of the drugs.

Providing inconvenient facts is not simply a possible role of a scientific advisor or teacher. Weber views the scientist's task as compelling an audience to accustom itself to the existence of such facts. He even calls it a moral achievement of advisors if they succeed in doing so (Weber 1948a: 147). A scientific teacher, Weber maintains, "stands in the service of moral forces; he fulfils the duty of bringing about self-clarification and a sense of responsibility". That aim is also served if scientists provide concepts and frameworks that define and organize diffuse phenomena into clear and consistent wholes. They can introduce, for instance, ideal-types and models for this purpose. By developing and using these scientific techniques of systemization, abstract, and complicated items such as 'bureaucracy' or 'syndicalism' can be practically addressed. Organizing complicated phenomena in a simple and consistent conceptual form advances their study them and gives policymakers a better understanding of the policy issue under consideration (Weber 1948a: 151–152;1949a: 20, 24).

Summarizing, we may conclude that Weber builds his ideal of scientific neutrality on the logical distinction between factual statements and evaluative judgments. Intellectually honest scientists honour this distinction and refrain from value-judgments in their teaching and in scientific advice. That does not preclude them, however, from contributing to policymaking. They can systemize the diffuse concepts that are often involved in policymaking, and they can note the limitations and (unintended) consequences of policy instruments that are under consideration. It is the duty of Weberian scientific policy advisors, moreover, to present these findings to policymakers as inconvenient truths to make them reconsider their policy positions critically.

3 Some Inconvenient Observations

This Weberian position has the quality of simplicity and clarity—a clear fact-value distinction that parallels the difference between autonomous and politically neutral scientist-advisors on the one hand and policymakers in their political value sphere on the other hand. Examining Weber's position more closely, however, gives rise to a series of questions:

1. A particular aspect of Weber's own view is obviously not value-free. Weber notes what a scientist and advisor *should do* and what he *should not do*. The demand to refrain from certain value-judgments, and living up to that demand, itself involves employing certain values. Weber also mentions intellectual honesty and integrity and the moral task to enlighten policymakers about their blind spots. The question arises of how, exactly, such value-judgments and virtues relate to the ideal of ethical neutrality and value freedom.
2. As we have noticed, Weber notes that scientists are motivated by, and that scientific practice in general is guided by, the idea that this type of activity is valuable. Science as a vocation involves more than only having a (well-paid) job. In earlier days, scientists studied nature to honour God's greatness; today, they might want to make the world a better place, Weber acknowledges. Furthermore, he maintains that all scientific disciplines are guided by some understanding of what is worth knowing (e.g., medicine has life and the

reduction of suffering as its core values, and cultural studies have the value of existing cultural phenomena). Weber does not appear to think that these observations are at odds with his ideal of ethical neutrality or value-neutrality or the distinction between science and politics. However, can this separation really be maintained? The fact that cultural sciences such as sociology or political science must value cultural phenomena as such and think it worthwhile to study them might not bring these sciences into the field of politics. They must decide, however, what particular phenomena they want to investigate. Should they study, for instance, the influence of gender on working routines in governmental bureaucracies? If they do, they might have research findings, possibly inconvenient facts that they otherwise would not have had. The evaluative choice of what is worthy of study determines on which issues and on what factors policymakers might be critically addressed. That statement appears to imply that the choices that scientists make, for instance in formulating research questions, actually do have consequences for policymaking. The issue again is, what do these consequences mean for Weber's elegant differentiation between science and politics?

A similar point concerns the clarifications of concepts that scientists can provide for policymakers. Is systemizing and simplifying vague and complicated concepts such as 'democracy' or 'effectiveness' or 'free market' or 'bureaucracy' merely a technical and strictly value-neutral matter? Weber himself presents a much-used ideal typical concept of 'bureaucracy' that, as he acknowledges himself, emphasizes the value of effectiveness. Building models and formulating ideal types appear to involve, therefore, making choices that prioritize certain values. Here, again, there appears to be reason to question the neat division between science and politics.

3. A third issue concerns Weber's stance on what a scientific policy advisor should do. Inconvenient facts should be presented to policymakers. However, what would that imply in practice? Does it mean that advisors should provide insights only when their input is requested? Should they actively try to obtain access to policymakers and confront them with their findings? Or should they take the role of a public activist and try to influence media and public opinion to make policymakers consider their findings? In short, can 'presenting inconvenient truths' be done neutrally? When scientists

decide how they wish to be involved in the political process, does this not imply making political judgments?

These observations impel us to investigate the ideal of value freedom in the position of the scientific policy advisor. Can ethical neutrality and the distinction between science and politics be maintained, as Weber posits? What, exactly, are the lessons we must draw from these complications? Should we set the ideal of value freedom aside and opt for a different one?

4 What to Expect in This Book

This critical overview of Weber's position on ethical neutrality leads to the following central question: *what type of value decisions should expert advisors make, particularly in the fields of Policy Analysis and Public Administration, and how can they address these issues?* To answer this question, several steps will be taken in this book. First, central terms must be clarified and defined, most importantly, 'value' and 'valuation'. In addition, an overview of types of decisions and types of values will be given (Chap. 2). The next chapters are concerned with the challenges that a position of ethical neutrality in policy advice, such as the one Weber presents, must face. Here, the inconvenient observations that were made above will be further elaborated (Chaps. 3, 4, and 5). It will be established that scientists inevitably must address value-judgments (Chap. 6). In the following set of chapters, the focus is on how a scientist might address these challenges. A variety of answers is distinguished and assessed (Chaps. 7, 8, 9, and 10). In the concluding chapter, the findings of this survey of arguments are then combined.

Now more than ever, a concern for the diminishing status of scientists and sciences in general can be witnessed. In particular, the role of scientific policy advisors is increasingly questioned—not only in politics but also or even particularly so in public opinion. Scientific claims on climate change and its causes are questioned. Some people express serious doubts about scientific claims on the effectiveness and safety of vaccines. In addition, the political neutrality and disinterestedness of economists and other social scientists giving policy advice are not taken for granted, particularly when their expertise is in areas of political controversy. In critiques such as these, it is suggested that scientists in fact serve some special interest (of pharmaceutical companies, for instance) or a political partisan ideal (socialist, environmentalist, or any other). A reader expecting in this book a study of the causes and consequences of this growing distrust of science, or even a remedy, will be disappointed. It does not go into such empirical questions. That does not mean, however, that the investigation that is performed here is of no relevance to that topic. It, in fact, addresses an issue that is basic to the contemporary discussion. Most of those who criticize current practices in scientific advice appear to rely on the same ideal as do those who are concerned about science's diminishing status. Both parties presuppose an ideal of the ethically neutral scientist. For both positions, it is crucial that the distinction between science and politics can somehow be upheld. Thus, feasible answers to both of the challenges mentioned above are essential.

Scientific advisors, of course, work in all types of institutional settings. Some are employed in universities, some in privately owned consultancies or in think tanks, and others in official advisory bodies. All encounter in their roles as expert advisors particular problems; the tension between research and politics for each will imply particular challenges. This book does not address this diversity of typical problems. Rather, it focusses on an issue that is fundamental for all of them.

A final note on the place of Weber in this book. In this chapter, Weber's work proved very helpful in presenting a picture of the ideal of ethical neutrality in scientific policy advice and of some of the questions that accompany that ideal. We could further investigate the issue of value-freedom in science by delving deeper into Weber's argument of the fact-value-distinction or by analysing its relationship to his distinction between the scientific and the political value spheres (Weber 1948b. For such an investigation see, for instance, Strauss 1963: chap. 2). This book, however, does not intend to be a textbook or a critique of Weber. It therefore pursues a different path. However, Weber's example here is not a randomly chosen position, and we will regularly return to it.

References

Die Zeit. (2016, July 14). Hört auf die Experten! Retrieved January 9, 2017, from http://www.zeit.de/2016/30/fachwissen-experten-schwierige-zeiten-loesungen.

Fischer, F. (1980). *Politics, values, and public policy: The problem of methodology*. Boulder, CO: Westview Press.

NRC Handelsblad. (2015, April 11). De feiten zijn geheel naar wens. Retrieved January 9, 2017, from https://www.nrc.nl/nieuws/2015/04/11/de-feiten-zijn-geheel-naar-wens-1483345-a1238207.

NRC Handelsblad. (2016a, August 20). Nee meneer de politicus, mijn statistieken zijn niet links. Retrieved January 9, 2017, from https://www.nrc.nl/nieuws/2016/08/18/nee-meneer-de-politicus-mijn-statistieken-zijn-niet-links-3838884-a1516993.

NRC Handelsblad. (2016b, October 7). Oorlog in de Islamologie. Retrieved January 7, 2017, from https://www.nrc.nl/nieuws/2016/10/07/oorlog-in-de-islamologie-haters-tegen-wegkijkers-4682112-a1525595.

Strauss, L. (1963). Natural Right and History. The University of Chicago Press (London).

Weber, M. (1948a). Science as a vocation. In H. H. Gerth & C. W. Mills (Eds.), *From Max Weber. Essays in sociology* (pp. 129–156). London: Routledge & Kegan Paul.

Weber, M. (1948b). Religious rejections of the world and their directions. In H. H. Gerth & C. W. Mills (Eds.), *From Max Weber. Essays in sociology* (pp. 323–359). London: Routledge & Kegan Paul.

Weber, M. (1949a). The meaning of 'Ethical Neutrality' in sociology and economy. In: E. A. Shils & H. A. Finch (Eds.), *The methodology of the social sciences* (pp. 1–47). Glencoe, ILL: Free Press.

Weber, M. (1949b). 'Objectivity' in social science and policy. In: E. A. Shils & H. A. Finch (Eds.), *The methodology of the social sciences* (pp. 50–112). Glencoe, ILL: Free Press.

CHAPTER 2

Setting the Stage: Epistemological and Social Values for Scientific Research and Advice

Abstract Advocates of neutrality (or value freedom) in science and scientific policy advice distinguish between two types of values. Epistemological values, on the one hand, are closely linked to an epistemological ideal of arriving at true statements; they encompass the methods, rules and virtues that lead to well-tested theories or convincing interpretations. Social values, on the other hand, concern ethical, cultural and political values. A sophisticated understanding of neutrality in scientific advice maintains, against feminists and other critiques, that social values might play some role, but that the core decisions in science can and should be purely guided by epistemological values.

Keywords Epistemological values · Social values · Method

1 Introduction

Max Weber links the ideal of ethical and political neutrality in scientific advice to the distinction between facts and values. Promoting certain values and making evaluative judgments belongs, he maintains, to the domain of policymakers, not that of scientists. A scientist can only provide arguments based on facts. Notwithstanding such claims, Weber also notes values that a scientist should maintain in his research and advice. Remarkably, even for an advocate of neutrality such as Weber, science is

B. Tholen, *Virtue and Responsibility in Policy Research and Advice*,
DOI 10.1007/978-3-319-65253-5_2

not and cannot be completely value-free. Here, a contradiction looms—'values for a value-free science'.

However, there might be an elegant solution to this paradoxical situation. The threat of a contradiction can be removed if two types of values are distinguished—either values that have a proper place in science and are strictly limited to matters of scientific method, or social or political values for which scientific practice is off limits. In this chapter, such a dichotomy will be introduced, elaborated, and discussed. It will prove helpful to present the different challenges to the ideal of ethical neutrality in scientific advice in the following chapters. First, however, a few words on the notion of values.

2 Values, Preferences, Interests

Values, ethicists emphasize, should not be confused with emotions or interests. Of course, acting in accordance with (or contrary to) the values that one holds might cause joy and satisfaction (or pain and discomfort). However, the term values predicates something that cannot be reduced or equated to these emotions. Having a good feeling about something does not mean in itself that the act was good (or bad). Nor can values be identified with interests. To be sure, both can be understood as motivations for certain acts—acting in a certain way because one thinks it a good thing, and acting thus because one judges doing so to be to one's advantage. Occasionally, these two considerations can have a similar result. However, these motivations are different in quality. When values are involved, we would expect people to be able to justify their actions by referring to something that is worthwhile or of importance beyond their personal needs and preferences. Anyone who identifies interest-realization and acting in line with one's values confuses 'I want that' with 'that is a good thing to do' (MacIntyre 1985; Taylor 1985: 85).

Values often are distinguished from norms, rules, and principles. Thus, values are occasionally considered aims that people deem worthwhile, whilst norms or rules are viewed as the instruments to realize them. In other cases, rules or norms are regarded as the limits for value-realizing behaviour; they mark off types of actions that are prohibited or obligatory. In this book, the word 'values' is used as a general term that surpasses such distinctions and definitions. Only in the last chapters, when addressing different ethical traditions, will evaluative terms such as rules or virtues refer to more-specific content.

Weber distinguishes values from facts. He connects the former to 'what ought to be done' and the latter to 'what is'. When he addresses the issue of value-freedom, Weber simply relates facts to observation and empirical investigation. He does not elaborate on what we can actually mean by 'fact' or whether a fact can simply be observed. For the moment, we will leave it at this simple indication. In later chapters, we will find, however, how observation and the determination of facts are relevant for the types of judgments scientific advisors must make.

3 Two Types of Values

In many elaborated advocacies for a neutral science, it is acknowledged that the type of value-free science that is intended is not completely void of values. Weber, for instance, distinguishes between two types of values. On the one hand, there are 'the rules of logic and method that are presupposed as valid in all scientific work'; on the other hand, he notes 'cultural values' (Weber 1948: 143). Others present a similar distinction, although different terms are often coined. They distinguish, for example, between scientific and extra-scientific values, cognitive and non-cognitive values, epistemological (or epistemic) and societal values or between methodological and contextual values (e.g., Lacey 2005; Popper 1976; Douglas 2009). These advocates maintain that the first type of values—here the term *epistemological values* will be used—concerns the rules, virtues, and norms that guide good science. Following epistemological values is what good or proper scientists do. Doing so ensures that they find the truth and offer the type of advice that ideal scientists can attain. Living up to these rules, values and virtues enables them to distinguish better theories from worse ones about causal relationships in phenomena in the observable world and to distinguish better interpretations from worse ones of the meanings that certain phenomena have for people.

Epistemological values are closely linked to normative theories of science; from these values, particular methodological guidelines and rules are derived. In line with a Popperean falsificationist ideal of science, for instance, these guidelines and rules involve a commitment to systematic testing of theories, including an attentiveness to consistency in theories, logical deduction of hypotheses, and exposition of testing conditions. From Lakatos' sophisticated extension of this approach, additional concerns follow such as theoretical progression and an increasing empirical content (Lakatos 1973). In an often-cited article, Kuhn suggests a list of basic scientific values for addressing theories, including accuracy, consistency, scope, simplicity, and fertility (Kuhn 1977). Epistemologies that focus on the interpretation of meanings entail analogous, although not always completely similar, rules and methods. Here, not only the reduction of puzzlement but also robustness and rigour in data gathering and analysis are demanded to minimize bias. Minimizing bias implies generalizability, transferability and replicability (e.g., Bailey 1992). The epistemological values include the rules and guidelines that all students all over the world must learn.

Advocates of scientific neutrality differ on the question of which values exactly should be included in the list of epistemological values. This disagreement largely follows discussions within epistemology, which can be seen most clearly in the social sciences in disputes between adherents of an explanatory science that is focussed on general theories and defenders of an interpretative science that offers assessments of the meaning of cultural phenomena. These disagreements, however, need not undermine the very idea of a distinction between epistemological and social values. The discussion, or at least part of it, may be understood as a proof of the ongoing search within science for rules and methods that guarantee unbiased conclusions and advice. All participants intend to refine further the scientific practice by offering additional guidelines for a neutral stand.

Social values are quite different. They express social, political, and ethical ideals. They include individual attitudes and behaviour such as friendliness and ambition or responsibility and authenticity. They also include societal ideals such as democracy or theocracy, family values or individualism, and merit or gender equality. Societies encompass their own set of values, often quite different from the values to which members of other societies adhere. Additionally, within societies, the presence of a variety of values is valid. Weber presents this social and political value pluralism in his analogy of the warring of gods under polytheism (Weber 1948).

The advocates of neutrality maintain that science cannot determine which social or political value should have priority and that this type of value has no place in scientific practice and advice. In fact, many of the epistemological rules and guidelines are aimed at keeping them out. These include, for example, methods that aim to minimize bias in case-selection, in interview questions, in coding data, and in analysing statistical material. Epistemological values are, therefore, at odds with societal values. They exclude each other in scientific practice and advice.

Epistemological values, to be sure, are not the single privilege of science. Many rules and values that guide science in finding better interpretations and explanations of the observable world also function in daily life. There is no categorical difference between scientific knowing and common sense, or at least there need not be such a difference (Lacey 2005; Popper 1976). The only privilege that science, at least a well-functioning science, has is the ability and opportunity to try to live up fully to these values. Its knowledge claims are guided by epistemological standards and not by social values. It is precisely for this reason, advocates of scientific neutrality maintain, that science can provide expert advice to policymakers. By relying only on epistemological values, science can provide truth to power.

4 Challenges to the Distinction Between Epistemological and Social Values

The distinction between two types of values, epistemological values and those that express social and political ideals, creates an attractive solution to the first inconvenient observation made in the first chapter. It opens up a path for a more sophisticated ideal of value-free science, a science that is neutral in the sense that it is free of social values.

However, if the advocacy for a neutral science comes to rely on this distinction of values, further critical questions arise. A first question is analytical: is the distinction between epistemological and social values actually tenable? Can there be pure epistemological values, completely free from social ideals? The second question takes the possibility of an analytical distinction between the two types of questions as a given. It asks, is a science that does not (also) employ social values actually possible?

The possibility of distinguishing between epistemological and social values might be questioned in several ways. One might initially note that ideals and values about how to do science well have changed over time and in different contexts. Only from the thirteenth century onwards have researchers begun to use experiments, and centuries passed before rules about repeated observation were formulated. Ideas and accompanying rules and values about using experimental groups and test groups and about double blind tests came much later. The same is true of guidelines for formulating testable hypotheses and the introduction of statistical methods and measures of statistical significance. Critics might even emphasize that epistemological ideas about knowledge and truth changed over time; they could refer to a shift from a reliance on a Cartesian rationalism in early modern days, via the empiricism of Locke and others, to a form of critical rationalism along the lines of Kant and Popper. They could also note that certain methods, rules and ideals of knowledge can be linked to the societal developments and social values of the time of their invention. Certain ideals of truth and values of research are closely connected to the broader social values of the Enlightenment, for instance. The introduction of statistical methods, to provide yet another example, was linked to the development of strong centralized state powers. For example, it helped organize large armies, facilitated efficient taxation, and made uniform state-education possible. In short, statistical methods connected to certain values of societal organization and control (e.g., Foucault 1975).

Following this historical line of thought, a critic might also focus on the individual level. He could provide examples of methodological methods and guidelines that have been developed by scientists that acted upon certain social or political ideals (e.g., Latour 1987; Shapin and Schaffer 2011).

These historical observations might certainly be accurate. However, they do not necessarily undermine the validity or feasibility of the distinction. For example, all of the methodological elements that have been introduced in science since the Middle Ages can be understood as steps directed at realizing the same value—eliminating error and bias. The same is true for the shifts in epistemological guidelines; they may be understood as steps in an ongoing effort to elaborate what we mean by 'truth' and 'true claims about the observable world'. Moreover, specific scientists having designed certain methods to prove a certain point does not make that method an expression of social value; for that matter, nor does doing so undermine the point's epistemological value. An advocate of scientific neutrality would maintain that the epistemological value of a statistical test is not in whether it can help to support one's belief. Rather, it is in whether it can help determine whether that belief is true. Telling the history of the discovery of certain methods and the ideas behind them is categorically different from expressing the epistemological validity of these rules and values (Lacey 2005).

Another means of questioning the distinction between epistemological and social values digs deeper. It does not focus on the history of epistemological values but rather disputes their socially neutral character head-on. This type of comment has been presented in critical studies such as cultural examinations of feminism. Longino, for example, maintains that feminist researchers employ values in their judgments of empirical theories that differ from those that are commonly used. Relying on Kuhn (1977), she claims that mainstream research is guided by values such as accuracy, consistency, scope, simplicity, and fertility in its effort to find and test empirical theories. The work of feminist studies, however, shows a reliance on values such as empirical adequacy, novelty, ontological heterogeneity, complexity, mutuality of interaction, application to current human needs, and diffusion or decentralization of power (Longino 1995: 385). This difference, according to Longino, must *not* be understood as expressing fundamental differences in how men and women understand the world (because of differences in biology or socialization). Such an understanding is not really convincing, Longino argues, given what various male and female scientists actually do. The difference between mainstream and feminist research expresses something else—a different orientation on the value of knowledge. In feminine approaches, doing research is clearly guided by a particular political value—revealing (the workings of) gender relationships in all types of social settings and practices. This concern is expressed by seeking novel theories rather than trying to find those that are consistent with established ones, or searching for understandings that dig more deeply and show complex underlying power relationships rather than concentrating on broadly applicable simple ones. Mainstream epistemological values do not guide the researcher in discovering dominating gender relationships; consequently, they express support for the status quo on societal power relationships (Longino 1995: 391).

This type of comment provides a more serious challenge to the ideal of a neutral science. Epistemological values might appear neutral in relation to social and political ideals, but actually they are not. Doing science is always political in consequence, and a scientist should be aware of this fact and act upon it. Even the mainstream scientist who claims to rely strictly on his rules of method is in truth politically involved when doing so. However, this fundamental critique of the distinction between epistemological values and social values does not necessarily mean that a neutral or sophisticated ideal of value-free science must be abandoned.

Advocates of scientific neutrality or value-freedom might grant that differences in values among researchers, such as the one feminists note, can be found. They might even accept that these values include ones other than only those of revealing (or not revealing) dominating gender relationships: for instance democracy, a sustainable environment, health, or any other social value. Furthermore, they might accept that the work of scientists is social-value guided in the sense that feminists note, even when the scientists themselves have no conscious intention of expressing and realizing a particular social value and think that their work is completely neutral. Advocates of neutrality can hold on to their ideal by granting non-epistemological values a particular, limited place in science. They can maintain that social values might be relevant and even unavoidable in some parts of scientific practice but that they are not in the core of it. In the areas that really matter, the epistemological values must rule in bringing about the truth.

Advocates of neutrality in science who adopt this approach acknowledge that social values can have a role, for instance, in the selection of research topics or in the formulation of research questions. However, in the core of scientific study—in data analyses, theory testing, and the like—there is no room for social values. Lacey, for instance, replies to Longino's critique of value neutralism by noting that the typical feminist values she mentions have a role only in the very first phase of research. After they have had their due in the selection of the topic, the neutral epistemological values kick in for testing the validity of theories (Lacey 2005: 218). Another champion of neutrality aptly referred to the phase in which social values might have a role as 'the *antichambre* of science' (Dahrendorf 1968). Weber follows the same strategy, we can now see, when he refers to the values of doing science and of what counts in different disciplines as 'worth knowing'. These values clearly are not epistemological in nature, but their relevance is limited to selection of research topics and questions. It is the rules of logic and method that are guiding science proper (Weber 1948: 143).

By following this strategy, advocates of scientific neutrality can uphold their answer to the first question this section started with; the distinction between social values and epistemological values is tenable. The strategy also suggests an answer to the second question. At least in the selection of research topics and research questions, social values might be inevitable.

5 What Is Next?

Through granting social values a particular albeit limited place in scientific practice, the ideal of a neutral science might be rescued. However, from the sketch given so far, this ideal remains rather vague. At least two issues must be clarified. The first concerns the exact role of social values in research. The metaphor of an *antichambre* suggests a clear distinction between some preliminary actions and the real scientific work. However, exactly what choices or judgments in a study can and must be made by epistemological values alone, and when can or must social values have a role? Are these latter values only relevant in the selection of topic and research question, or are they also involved in theory selection? What about the further phases? Do non-epistemological values have no necessary or inevitable role in data gathering and analysis, conclusion, and advice? The next three chapters offer answers to these questions. For convenience, the practice of doing research is roughly distinguished into three phases. First is the initial phase, which most likely comes close to what Dahrendorf has labelled the antechamber. It encompasses the selection of topics, research questions and theoretical approach (Chap. 3). The 'scientific core' of data gathering, analyses, and theory testing forms the second phase (Chap. 4). The third phase concerns drawing the conclusion and giving scientific advice (Chap. 5).

The strategy of distinguishing epistemological and social values and granting the latter a limited place brings up a second issue that needs clarification; that is, what should a scientist do when confronted by value-choices that cannot be answered by epistemic values (alone). Scientists are well trained in employing epistemological values—for example, the use of logic, guidelines for valid data gathering, the appropriateness of certain statistical tests, and rules for selecting the better theory. However, what about those decisions that inevitably involve social or political evaluations? Are there proper, even scientific, ways to approach such decisions? Answers addressing this issue will be the subject of Chaps. 7, 8, 9, and 10.

References

Bailey, M. T. (1992). Do physicists use case studies? Thoughts on public administration research. *Public Administration Review, 52*(1), 47–54.

Dahrendorf, R. (1968). Values and social science. In R. Dahrendorf (Ed.), *Essays in the theory of society*. Stanford: Stanford University Press.

Douglas, H .E. (2009). *Science, policy, and the value-free ideal*. Pittsburgh, PA: University of Pittsburgh Press.

Foucault, M. (1975). *Discipline and punish: The birth of the prison.* New York: Random House.

Kuhn, Th. (1977). Objectivity, value judgment, and theory choice. In Th. Kuhn (Ed.), *The essential tension* (pp. 320–329). Chicago: Chicago University Press.

Lacey, H. (2005). *Is science value free? Values and scientific understanding.* London: Routledge.

Lakatos, I. (1973). Falsification and the ethodology of scientific research programmes. In I. Lakatos & A. Musgrave (Eds.), *Criticism and the growth of knowledge* (pp. 91–138). Cambridge: Cambridge University Press.

Latour, B. (1987). *Science in action.* Cambridge, MA: Harvard University Press.

Longino, H. E. (1995). Gender, politics and the theoretical virtues. *Synthese, 104,* 383–397.

MacIntyre, A. (1985). After Virtue: A Study in Moral Theory. University of Notre Dame Press (Notre Dame, In.).

Popper, K. (1976) The logic of the social sciences. In G. Adey & D. Frisby (Eds.), *The Positivist dispute in German sociology* (pp. 87–104). London: Heinemann.

Shapin, S., & Schaffer, S. (2011). *Leviathan and the air-pump: Hobbes, Boyle, and the experimental life.* Princeton, NJ: Princeton University Press.

Taylor, Ch. (1985). Neutrality in political science. In Ch. Taylor (Ed.), *Philosophy and the human sciences. Philosophical papers 2.* Cambridge: Cambridge University Press.

Weber, M. (1948). Science as a vocation. In H. H. Gerth & C. W. Mills (Eds.), *From Max Weber. Essays in sociology* (pp. 129–156). London: Routledge & Kegan Paul.

CHAPTER 3

Deciding What Is Worth Knowing

Abstract In the first phase of a scientific study, a set of choices must be made for which scientific advisors cannot completely rely on epistemological values: doing scientific research, choice of research topic, research question and theories. The choices that are made in this phase of a study have implications for the advice that can be given. They determine about what issues inconvenient truths might be provided. Theory choices also lead to a particular value slope or appreciative system built-in in the conclusion.

Keywords Worth knowing · Research topic · Research question

1 Introduction

This chapter concentrates on the first phase of doing research—the choices of topic, research question, and scientific approach. It also involves the problem of initial engagement in scientific study. The first section addresses the following issue: to what extent can such choices be made relying on epistemological values alone? The next section addresses the consequences of these choices. The question of the last section is, what do these findings mean for the ideal of neutrality in science.

B. Tholen, *Virtue and Responsibility in Policy Research and Advice*,
DOI 10.1007/978-3-319-65253-5_3

2 Choices the Scientist Must Make

Becoming Involved in Scientific Research

The issue of why one should be engaged in science is a matter that Weber addresses head-on in *Science as a Vocation*. After acknowledging that doing scientific research is a means of making a living and having a career, he switches to a more philosophical vein. What might be the value of doing scientific research, of finding scientific truth? Weber's reply is that science presupposes an answer to this question but that it cannot itself provide one. The justification for doing science cannot but lie in an ethical (or theological or political) stance. As an illustration, he cites Swammerdam, who in early modern times searched for knowledge to show the path to God: 'Here I bring you proof of God's providence in the anatomy of a louse' (Weber 1948a: 142). Weber holds on to his role as an empirical researcher and merely notes that science presupposes a foundation in social or ethical values; he does not intend to provide such a foundation himself.

Karl Popper makes a similar point about the need for an ethical foundation for being engaged in science. He argues that the issue of whether one should try to acquire knowledge by a method of trial and error or should settle for some dogma or belief cannot be solved rationally or scientifically. Such a decision, he maintains, may only be made on moral grounds. He contends that any attempt to offer a rational argument for Critical rationalism would find itself in a circle or an infinite regression (Popper 1993: 232, 238). In contrast to Weber, Popper actually does provide (the outlines) of a normative theory underlying science. He insists that scientific knowledge could contribute to a better world, one that is not attainable if one dogmatically accepts some idea about what the world is like and how it functions. In a world in which policymakers were informed by the knowledge of scientists, there would be less wrongdoing, injustice, exploitation, and avoidable suffering such as poverty and unemployment. The role of the (social) scientist is to provide knowledge for policies that can systematically fight such wrongs and injustices. They must provide the input for a practice of social engineering—that is, *piecemeal* engineering by probing (limited) policies and dismissing them when they fail. Popper presents this technique in contrast to the dogmatic approaches to (large scale or holistic) policymaking and planning that he witnessed in authoritarian states (Popper 2002: 38, 84).

The Values of Scientific Disciplines

Within the field of Public Administration, policy studies, and management sciences, the idea that scientific research does have a task in furthering good government and social justice is broadly shared. The journal *Policy Sciences*, for instance, has as its motto 'Integrating knowledge and practice to advance human dignity for all'. Dunn, in his well-known and widely used handbook on policy research, asserts that the aim of this type of enquiry is 'to create, critically assess, and communicate information that is useful in understanding and improving policies' (Dunn 2012: 1).

This disciplinary value-outlook also is well recognized by Weber. He maintains that no scientific discipline is free from presuppositions, nor can it be. Each discipline needs some orientation towards that which is 'worth knowing'. Each discipline contains basic guiding values that give it its meaning. The medical sciences concentrate on what we can do to maintain life and diminish suffering. The natural sciences answer the question of what we should do if we wish to master life technically. Juridical thought shows when certain legal rules and methods of interpretation are binding. The historical and social sciences show how political, artistic, literary, and social phenomena can be understood and interpreted and how they came into existence. These disciplines thus presuppose that life is valuable and worth living, that it is a good thing to master nature, and that it is good to have law and to have knowledge of the meaning of cultural phenomena. However, the disciplines cannot provide scientific grounding to these suppositions (Weber 1948a).

Choosing the Research Topic and Research Question

The disciplines have their specific presuppositions on what is worth knowing. These value-presuppositions, however, have a very general character. 'Furthering good government' or 'diminishing human suffering' is rather vague—too vague to guide a researcher to any particular research question. For example, which diseases should be studied, what types of (and whose) sufferings should receive priority in research, and what aspect of governance deserves scholarly attention? (One might argue that these disciplinary guidelines are clear, at least, in one respect—in the types of research they exclude. Research into effective methods of torture, the possibility of building a 'doomsday machine', or the means of effective dictatorship evidently will not do.).

Can epistemological values guide the selection of research topics, questions, and theories? An authoritative perspective on epistemological guidelines for doing research and choosing theories has been presented by Imre Lakatos. Lakatos adopts Popper's scientific ideal of falsificationism, an ideal that essentially demands that empirical theories must be tested by systematic observation. Lakatos elaborates this falsificationism on two important points. First, he maintains that it would be reckless to abandon a theory whenever one encounters an observation that does not fit into it. There might be something wrong with the observation or some presuppositions about the observation. Lakatos, therefore, argues for a more sophisticated type of falsificationism; a theory should be abandoned only if there is an alternative. One should abandon the original and switch to the rival if the latter suits the observations better. To be more precise, the new hypothesis should survive the empirical tests that the old one cannot, it should be able to explain why the old theory had the success it had, and it should have a larger scope (more empirical content). In line with these guidelines, for instance, one should prefer Einstein's system to that of Newton because it can show that Newton's idea of gravity only works for low velocities in particular circumstances. Lakotos' second elaboration concerns his idea of research programmes. Such a programme consists of scientists that share a particular research outlook. Within that common outlook, these scientists work on the improvement of theories, that is, revising older ones in the light of new observations. This practice is perfectly all right, according to Lakatos, as long as it leads to ever better theories according to the criteria just mentioned (Lakatos 1973). An example of such a research programme might be the ongoing effort of a group of scholars to find an ever-better explanation (and thus prediction) for the outcomes of bargaining practices starting from basic assumptions about individual rent-seeking and formal power-positions. Observations that are at odds with the current theory (anomalies) inspire researchers to redefine particular presuppositions slightly or to add a variable.

Although Lakotos' approach is only sketched briefly here, it clearly contains specific epistemological rules on how to proceed in topic and theory selection. To further scientific knowledge, one should propose theories that offer improvements compared with current ones—for instance, theories with a greater scope (more empirical content). Such a practice implies involvement in existing research programmes and their specific thematic focus.

Lakatos' epistemology clearly provides some guidance. However, it does not completely determine the choices a scientist must make. It does not tell scientific advisors to which research programme they should attach themselves if they wish to work on a particular societal topic. For instance, they might also study bargaining processes from a completely different perspective. They might be interested in the effects of bargaining on the long-term relationship between the actors involved. Alternatively, they might ask whether bargaining is more cost-efficient than some other means of regulating, or they might want to know what it means for civic trust. Within a particular research programme, Lakatos' guidelines also leave many options open. His epistemology calls for theories with more empirical content, but it does not tell scientists in which direction they should enlarge the scope. Should students of bargaining outcomes, for instance, try to expand their theoretical explanation to the European Union (EU) and other international fields, or should they try to include micro-decision-making between colleagues at the work-floor level? The epistemological guidelines do not indicate which further variable they should try to include in the model when anomalies start to accumulate.

Sociologists and historians of science have noted that social conditions, prevailing societal ideas, but also work-floor conditions in laboratories or how research is funded can be considered to have influence on choices that individual scientists make. Research topics are often chosen that follow prevailing political concerns or fashionable topics. Scholars focus on issues that promise funding or promotion, or they simply follow the lead of others (e.g., Kuhn 1996, Latour 1987, Shapin and Schaffer 2011). For sociologists, these findings are interesting because they explain empirically the behaviour of scientists. For our investigation, these observations are relevant in another way. They illustrate the field of choice that scientists have when they remain within the limits that the rules of epistemology grant them. The social values (exemplified in the choices that the sociologists brought to light) fill the void that is left when epistemological rules and methods had their due. The fact that social circumstances tend to steer the choices of scientists in certain directions might be considered a problem (I will address this issue later). However, it does not contradict the conclusion that epistemological values alone are not sufficient in this phase or research.

3 Impact of Research Choices

Thus far, the focus in scientific choices has been on the input side; it was asked on what grounds the choices can be made. The view now will be on the output side; what does the choice of particular research topics, questions, and theories entail?

A first observation is that the selection of research topics will determine on which subject knowledge will be developed and, thus, in Weber's terms, inconvenient truth can be provided. The same is true for theory selection and determination of relevant variables. Consider, for instance, the variety in impact of two studies on the effects of the use of particular policy instruments. One study employs a theory that distinguishes between different types of communication. It might provide findings that help a communications officer to obtain better results. Another study that not only focusses on communication but also considers other policy instruments might identify a truth that is inconvenient to the communications officer altogether—that using a stick or a carrot is far more effective than a sermon in whatever form. Another example is the research into the functioning of a particular type of governmental agency. One might choose a theory that distinguishes between different operating procedures that such an agency can employ and relates them to effectiveness. One might also opt for a theory that distinguishes between different types of agencies and links these to efficiency. Scientific advisors who chooses the first approach might obtain findings that enable them to comment on certain policies of an agency. Those who takes the second approach might obtain results that question the existence of the agency.

There is a further point to this selectivity of theories, an issue that is emphasized, among others, by Charles Taylor. Empirical theories, Taylor maintains, set the crucial dimensions through which phenomena can vary. A theory might allow, for instance, that a certain type of agency employs different types of policy instruments, the one more effective than the other; or a theory might allow that public institutions can be more or less able to address social conflict. Although these theories explicitly allow variation on certain issues, they imply on others a given and evident reality. The existence and value of the agencies that can employ different measures appears as a given. Similarly, it is presupposed in the other example that in every society, some amount of social conflict is normal. From the perspective of these theories, the possibility that this type of agency has no value at all or that a society might be without conflict cannot exist. Similarly, does accepting the theory that free trade contributes to national welfare (or the theory that says that is does not, for that matter) imply that national welfare is of value? When we accept an explanatory theory, we accept the implicit valuation of what can count as good and normal—for example, good and normal policy, agency, behaviour, and political system. Theories, Taylor maintains, have a built-in value slope. Theories are frameworks of explanation that do not explicitly argue for particular values; they do not utterly bind us to them but subtly lead us to them. Acceptance of a certain explanatory theory in fact means that the onus of normative argument is distributed in a certain way (Taylor 1985).

A similar point is made by Vickers concerning policy advice. Research for policies inevitably is selective in how it approaches a policy issue. Consequently, a policy report always impresses upon the reader a readiness 'to distinguish some aspects of the situation rather than others and to classify and value these in this way rather than in that' (Vickers 1995: 82). He calls this phenomenon the 'appreciative setting' or 'appreciative system' of every scientific policy advice (idem 69–82). A study, for example, on the possibilities of using nudging-techniques to stimulate illegal aliens to leave the country suggests that it is proper to stimulate the *sans-papiers* to leave and that there is nothing wrong about trying to do so with socio-psychological techniques.

In the field of Public Administration and Policy Studies, surveys of power and influence provide typical examples of the effect of research choices. How the researcher decides to measure power can have a severe effect on the outcomes. If one decides to focus on formal positions and decision-making authority, one might identify completely different power holders than one might have found in a network analysis or a process analysis. In one study, the Dutch king would be designated as very powerful (being a formal member of government); in the other, his name would be much lower on the list.

A further reason why power research deserves particular attention here is the effect of moving power relationships into the light. Of course, publication of any type of knowledge might have its effect. Research findings might become news and set the agenda in public discourse. Even more elementary is that by giving research attention to some subject (e.g., a group of people, a mental disorder, or pollution), that subject is highlighted as something that is worthy of study thenceforth in the public consciousness. However, in the case of power, more is occurring. Being in power and being able to retain power occasionally means that one must be viewed as being in power. That explains the use of sceptres, thrones, uniforms, and statues (Schmitt 1954). Research shows that these power holders are of service to those in power. Things are different in power relationships that thrive in the dark, unnoticed by the public eye. Exposure of power relationships behind the scenes might affect the position of those power holders, for instance because citizens start demanding investigations and transparency. Thus, the choices made in power research might easily become choices about supporting or tempting existing power positions.

4 Summary

In this section, the focus was on the first phase of scientific advice—the choice to do research and the decision on what to study. In science, and in research for policy advice, we are not simply seeking any truth but rather interesting truth. We are interested in knowledge about things that we care about, that we consider worthwhile to know, that strikes us as significant. Science itself cannot answer the question 'what is worth knowing?' It is not even able to explain why it is worthwhile to obtain scientific knowledge at all.

The decisions that scientists and scientific advisors must make concerning subject of study and theory selection are, to a certain extent, issues of social values, not epistemological ones. In justifying them, one cannot but refer to what is socially valuable and what might be worth preserving. These choices seriously affect the scientific advice that can be provided. They determine on what issues inconvenient facts can be presented to policymakers. Furthermore, they result in advice that contains a particular 'appreciative system' and not another, thereby 'forcing' the onus of argument (on valuable ends and the set of feasible instruments or categories) upon a particular side in political and societal debate.

References

Dunn, W. N. (2012). *Public Policy Analysis.* London: Routledge.
Kuhn, Th. (1996). *The Structure of Scientific Revolutions.* Chicago: University of Chicago.
Lakatos, I. (1973). Falsification and the Methodology of Scientific Research Programmes. In: Lakatos, I. & A. Musgrave (Eds.), *Criticism and the Growth of Knowledge* (pp. 91–138). London: Cambridge University Press.

Latour, B. (1987). *Science in action: How to follow scientists and engineers through society.* Harvard University Press (Cambridge, Ma.).

Popper, K. (1993). *The Open Society and Its Enemies, Vol 2.* London: Routledge & Kegan Paul.

Popper, K. (2002). Science: Conjectures and Refutations. In idem, *Conjectures and Refutations* (pp. 43–78). London: Routledge.

Schmitt, C. (1954). *Gespräch über die Macht und Zugang zum Machthaber.* Pfullingen: Neske.

Shapin, S. & Schaffer, S. (2011). *Leviathan and the Air-pump: Hobbes, Boyle, and the Experimental Life.* Princeton, NJ: Princeton University Press.

Taylor, Ch. (1985). Neutrality in Political Science. In: Ch. Taylor (Ed.), *Philosophy and the Human Sciences. Philosophical Papers 2.* London: Cambridge University Press.

Vickers, G. (1995). *The Art of Judgment.* Sage (Thousand Oaks CA): A Study of Policy Making.

Weber, M. (1948a). Science as a Vocation. In: H. H. Gerth & C. W. Mills (Eds.) *From Max Weber. Essays in Sociology* (pp. 129–156). London: Routledge & Kegan Paul.

CHAPTER 4

Deciding on Tools and Risks

Abstract The choices that must be made in the second phase of a scientific study also inevitably involve social values. First, there are the values that urge the scientist to respect the subjects he researches (e.g., animals, environment, or human beings). Second, choices must be made between measuring instruments that are epistemologically equivalent but might have different social consequences. Third, choices must be made about the level of risk of being wrong that is acceptable.

Keywords Risk of being wrong · Method · Research ethics

1 Introduction

Most advocates of a value-free or neutral science acknowledge that in choice of topic and in theory selection, social values inevitably have a place. The neutrality (or objectivity as some prefer to call it) of science and scientific advice is guaranteed, they maintain, by the scientific method. Once the research question is formulated, only epistemological values rule. Deviance from the rules and methods of data gathering and data analysis makes for bad science. In this chapter, several challenges to this position will be presented. These comments do not question the relevance of the epistemological values and, thus, of the scientific method. However, they do entail that in this phase of science, non-epistemological values inevitably have a role. The first challenge is about the ethical

B. Tholen, *Virtue and Responsibility in Policy Research and Advice*,
DOI 10.1007/978-3-319-65253-5_4

limits of research. The choice of labels and categories and of measuring instruments is next. The final section focusses on deciding how much risk of being wrong one should accept.

2 Addressing Research Objects

Over the decades, research ethics have received much media coverage. A remarkable number of cases in which researchers clearly violated the rules of doing research have become public. Scholars have been accused, and found guilty, of fraud, curbing, tweaking, and even making up results. Research ethics, here, involve living up to the methodological rules of gathering data, analysing them, and using them to test theories or support particular interpretations of meaning. The relevance of this type of ethics for research is never disputed—it is understood as the core of doing science. This type of ethics expresses the epistemological values. It supports the claims to knowledge that science makes. Anyone who manipulates data, selectively uses them in the testing of theories and interpretations, or presents conclusions that the findings cannot support is not involved in finding the truth but rather in deceit. It makes for bad science. In fact, because the epistemological values are not honoured, it is not science.

There is a second type of research ethics, the relevance of which is commonly acknowledged. This type of ethics does not concern epistemological values but (a particular type of) social values. Doing empirical research often has direct consequences for the subjects under investigation. For example, experiments and other forms of data gathering can affect natural surroundings, plants, animals, the human body, individual wellbeing, and social trust. It is commonly accepted that certain values should be considered when researchers decide how and where they perform a study. For example, the welfare of test animals should be considered in drug tests; in surveys, personal information should be concealed to protect the privacy of the participants; and experimental treatments should not be employed without the consent of the patient. Concerning particulars, what this type of research ethics specifically demands is often not (immediately) clear. When, exactly, is an experiment degrading and disrespectful of persons? Is it justifiable, for instance, to let people punish others in some experiment (or let them believe that they do so)? Might a researcher, in some cases, lie to his respondents, for instance by withholding from them the true purpose of a study? This lack of clarity is a reason for ongoing disputes about whether particular types of study are acceptable. However, the relevance of social values on such matters itself is broadly accepted.

3 Categorizing and Labelling

The theories, models, ideal types, and frameworks that are chosen in the first phase of research must be further elaborated. Dependent and independent factors need further categorization, central terms must be defined, and theoretical terms must be made measurable. Therefore, further choices must be made; should, for instance, the gender factor be dichotomized as is often routinely done, or should one opt for a more specified categorization? Should a left-right distinction of political parties have one or more dimensions? Should 'civic participation' be understood in quantitative terms (number of contacts), or should the quality be measured (type of contacts or effect)?

The choices that one makes in this phase are, to some extent, comparable to the ones that have been made in the selection of research question and theory. They involve considerations in which epistemological values alone cannot do the job. Here again and more precisely, what is worth knowing must be considered. Again, the choices affect the type of (inconvenient) facts one will be able to present, and they affect the value slope or system of appreciation.

A word on classifying and labelling. The world as we encounter it is itself not labelled. It is we, the knowledge-developing species, who categorize, label, and even can be said to compose the categories. The same item or phenomenon might be labelled a means of transportation, a polluting device, a collector's item, a lethal instrument, and many things more. That does not imply, however, that we are free to 'construct' the world any way we want to in our labelling and categorization (see Boghossian 2006 for a clear comment on relativism). The point relevant for our argument is that we have options on what perspective we take on particular issues, which phenomena we choose to consider, and how we do so. These choices have consequences for the knowledge that can be provided later.

4 Choosing Instruments for Measurement

A further issue on data gathering concerns the choice of instruments for measurement. In every explanatory or evaluative study, how the relevant factors are to be measured must be decided; example metrics could be the level of welfare in a set of countries or the level of pollution, civic participation, or corruption. Such factors can often be measured differently. For instance, the level of carcinogenicity of pesticides might be measured by laboratory tests on animals, but the scientist could also rely on epidemiological studies. To measure the level of noise in the proximity of an airport, one could do a series of tests in the field, but you can also use mathematical models and simulate the effects of air traffic. In yet another case, one might choose between a Delphi study involving experts or a broad survey among a wide range of citizens to measure the effects of a particular policy measure against corruption. In all such cases, the methods for measuring can be sound in epistemological terms, each well-tested and leading to data of the same statistical quality. In such cases, for the choice between the methods of measurement, one must rely on other, social considerations.

This selection, which is based on non-epistemological considerations, has its effect, just like the theory choice in the last chapter has its value-slope effect. Two examples might illustrate the implications of these choices. In the 1970s, how to test for carcinogenicity of pesticides became an issue of dispute in the USA, although both existing methods (animal testing and epidemiological study) were considered epistemologically sound. Because an epidemiological approach would have made it more difficult to obtain a licence, it was not favoured by the producers of the pesticides (Majone 1989: 27). A similar issue was at stake in finding a standard for the level of noise acceptable in the proximity of a Dutch airport in the 1990s. One type of test relied on actual measurements in the field, the other on scenario studies in a mathematical model. Again, both approaches were sound in epistemological terms. However, if the modelling one were chosen, it would be easier for the airport to comply with noise regulations because deviations of routine flight behaviour would not occur in the model. Both examples illustrate what is at stake in the selection of mere instruments for measurement; selection determines whether there will be inconvenient facts of a particular type, and it influences the onus of argument in the political debate.

5 Addressing the Risk of Error

Science does not have the key to absolute truth or rock solid knowledge. This conclusion is the result of centuries of dispute over what observation and reason might provide. Neither empiricists, who rely on observation, nor rationalists, who trust reason alone, have proved convincing. Empiricism always relies on some form of induction, and thus on a limited set of observations. Rationalism must presuppose some access point to true knowledge, beyond the observable reality, a presupposition that attracts virtually no one (Popper 2002) any more . Most scientists currently accept the idea, and the consensus is followed in this book, that it is possible to distinguish better from worse interpretations of webs of meaning or causal theories by relying on systematic observation. This guiding idea leaves the possibility for progress in scientific knowledge without claiming that science at any particular moment has found the final and absolute truth about anything.

However, this fundamental attitude of doubt that is the core of modern science must not be confused with the ambition of science to limit error in the claims it makes. Errors can enter a study in every step of the research process. Researchers might make mistakes in identifying and measuring phenomena (for instance, by accidentally registering particular policies as fulfilling certain criteria when in fact they do not, by using datasets that are not sound, or by mistaking respondents for insiders when they are in reality outsiders). Something might go wrong in coding (for instance, interpreting certain words as indications of a positive stand when they are not). Moreover, in the analysis of data, mistakes can be made not only in the interpretation of qualitative data but also in statistical analysis.

Statistical analysis allows researchers to decide whether a theory should be rejected or accepted. In making that decision, they presume that their sample of data is a representative set. Because that might not be true, scientists might be wrong in accepting or rejecting a hypothesis following a statistical analysis. For this reason, a level of statistical significance is used in this type of analysis; it sets the level of risk of being wrong.

One might think it evident that a lower risk is always preferable; it allows one to make claims with less likelihood of being wrong. That, however, is too simple a picture, particularly when one considers that scientific knowledge is also meant to be used for policy advice. A first complication concerns the fact that decreasing one type of risk of being wrong often implies that the risk for another mistake becomes larger. This occurs, for instance, in statistical theory testing. Given the samples they have, researchers can make two types of error. Their tests might show that something is true when it is not. This type of mistake is called a 'false positive' or 'a fault of the first kind'. It can be compared with the alarm going off when there is no fire. The second type of fault (false negative) resembles the situation in which there is a fire, but the alarm did not go off. The test tells the scientist that something sought does not exist when it in fact does. If scientists want to diminish the chance of the first type of mistake (by choosing a stricter level of statistical significance), the risk of making a mistake of the second type increases, and vice versa. Suppose scientists use some test to determine whether a food additive contains a toxic substance. If they decide to set the level of significance at a stricter level, the test will deliver relatively few false positives and a higher number of false negatives. This setting will make the additive appear less dangerous than it actually is. Deciding to set laxer standards leads to more false positives and an increase in the chance of considering the substance toxic when it actually is not. A policy based on the first choice of the scientist might be setting legal limits for food additives too low, endangering public safety. The second choice, however, might bring about a policy of overregulation, affecting the profitability of the producing companies. Given such consequences, the decision on how much risk of error one is willing to accept comes down to the following question: which social values does one give more weight (Douglas 2007, 2009; Rudner 1953).

In social practice, there are often good reasons to choose either a smaller chance of a fault of the first kind or of the second. An anti-terrorist unit, for instance, might prefer to make a mistake of the first kind (false alarm) than miss a terrorist attack. Conversely, a court of law would rather make a mistake of the second type (releasing an actual criminal) than convict an innocent citizen.

Thus, in choosing between two given tests of the same rigour or between two types of mistakes, epistemological values might not be sufficient, and social values fill the void. Researchers can often evade such a decision. In the case of statistical analysis, for instance, they might also lower the level of risk by making their samples larger. The risk of both types of faults then decreases. Here, however, a second observation can be made. Excluding risks always comes at a price. Actions such as enlarging the dataset, upgrading measuring instruments, and increasing the levels of control in coding make the study more expensive, and more time is required to achieve results. Consequently, fewer funds are left for other issues that might be worth addressing. Furthermore, the ambition to increase the level of certainty might leave people waiting for knowledge that they would rather have earlier. Consider, for instance, some new medicine that might be a solution for patients that are seriously ill and for whom no other treatment is available. Should this drug be tested before introduction with the same rigour as food additives? Similar points might be made about, for instance, studies of social benefits for those worst off or the functioning of particular 'people processing organizations'. Occasionally, the need is not for the most robustly tested knowledge but rather 'true enough knowledge' or 'serviceable truth' (Elgin 2004; Jasanoff 1990, 2015), that is, 'knowledge that satisfies tests of scientific acceptability' but also 'supports reasoned decision-making' and does not sacrifice the interests of people in need or at risk 'on the altar of an impossible scientific certainty' (Jasanoff 1990: 250). This need eventually means that scientific advisors must find a middle course between taking too little risk (and having little to say) or too much risk (and producing ill-founded advice).

6 Summary

Even when one moves beyond its *antichambre* into the core of science, decisions that must be made by the researcher must consider social values. Because of the value we place on people, species, artefacts, and the planet, some types of experiments or tests are off limits. Here, social values place their mark. Furthermore, scientists must decide how much risk of error they accept. In setting the levels of risk, epistemological values alone are not sufficient. Scientists cannot but consider how serious a mistake would be; what is at stake, what is vulnerable, and what deserves our protection? Scientists must decide which instrument of measurement they will use. Epistemological values alone might not be sufficient to make this choice. Here, again, social values come into play.

The finding that social values have these roles in science at this stage does not imply that epistemological values are set aside. Moreover, it is certainly not true that non-epistemological concerns rule the actual decisions on rejecting and accepting theories. However, these non-epistemological values do influence the course and conclusions of research. Certain studies, that might help to test theories or interpretations of meaning that are now considered convincing, cannot be performed for ethical reasons. Certain outcomes might not end up in the conclusions because of the level of risk that was chosen. In addition, certain standards of safety and not others might be proposed, although all meet standards of scientific rigour.

References

Boghossian, P. (2006). *Fear of Knowledge*. Oxford: Oxford University Press.

Douglas, H. E. (2007). Rejecting the Ideal of Value-free Science. In H. Kincaid, J. Dupré, & A. Wylie (Eds.), *Value-Free Science?* (pp. 120–141). Ideals and Illusions. New York: Oxford University Press.

Douglas, H. E. (2009) *Science, Policy, and the Value-free Ideal*. Pittsburgh, PA: University of Pittsburgh Press.

Elgin, C. (2004). True Enough. *Philosophical. Issues, 14*, 113–131.

Jasanoff, S. (1990). *The Fifth Branch: Science Advisers as Policymakers*. Cambridge, MA: Harvard University Press.

Jasanoff, S. (2015). Serviceable Truths: Science for Action in Law and Policy. *Texas Law Review, 93*, 1723–1749.

Majone, G. (1989). *Evidence, Argument, and Persuasion in the Policy Process*. New Haven: Yale University Press.

Popper, K. (2002). Science: Conjectures and Refutations. In: idem, *Conjectures and Refutations* (pp. 43–78).London: Routledge.

Rudner. (1953). The Scientist *Qua* Scientist Makes Value Judgments. *Philosophy of Science, 20*,1–6.

CHAPTER 5

Deciding on the Advisory Role

Abstract In the third and last phase of scientific study and advice, decisions must be made on transforming the research findings into a message for policymakers. Such decisions involve selection of material. They also involve choosing between possible frames and communicative and rhetorical instruments. The selective role of the scientific advisor is larger when he is confronted with policy problems that are unstructured.

Keywords Rhetoric · Structuring problems · Advisory role

1 Introduction

In the first two phases of research, a range of decisions must be made that involve non-epistemic considerations. Before turning to the effect and importance of this practice, the third and last phase needs attention—giving advice based on scientific research.

Academic teachers and scientific advisors to politicians, according to Weber, should merely present the findings of their empirical studies. They should communicate their inconvenient facts to policymakers. However, what does this approach actually involve? What does 'presenting the research results' really encompass, what decisions are involved, and what type of values can guide those decisions? The first part of this chapter examines more closely the transformation of research findings into advice for policymakers. In the second part, the communication

B. Tholen, *Virtue and Responsibility in Policy Research and Advice*,
DOI 10.1007/978-3-319-65253-5_5

itself is at issue. It will be argued that in both transformation and communication, choices must be made that are not (completely) guided by epistemological values.

2 Simply the Facts?

The outcomes of scientific studies or the conclusions that scientists draw from them can be the basis of policy advice. Moreover, intellectual honesty, as Weber would call it, implies that this advice should actually express those findings. However, there is more to intellectual honesty than being honest with the facts. The research findings must be made meaningful to policymakers. This point is nicely illustrated by the common instruction to authors in journals in Public Administration and Policy Analysis to add 'points for practitioners' to their manuscripts.

Translating the Findings

A first aspect of making research findings meaningful is presenting them in a vocabulary that policymakers can understand. Theoretically, defined factors such as 'organizational resilience', 'social capital', or 'input legitimacy' must be translated, and that often implies making choices. The ordinary language term will seldom completely cover (or be limited to the field of) the theoretical term.

The vocabulary a scientist chooses, moreover, is hardly ever completely neutral. Words and concepts have positive or negative connotations. In cases such as freedom fighter/terrorist, this point is obvious; however, in other cases, the valuation involved in terms and phrasings is more subtle. The choice to formulate the advice in this or that particular wording thus means adding this or that value-colouring to the message. If scientists formulate the results of their studies in terms of public spending, they link their advice to a different frame than when they choose public investments, for instance. (One might say additional value-colouring because the chosen theoretical language also has its appreciative character or value slope. See Chap. 3). In the field of the natural sciences, words might be free of such value-colouring. The reason is most likely that they concern phenomena that we do not really care about (Dupré 2007). The social domain, the field that much of policy advice is concerned with, is precisely about issues that we value or abhor. The language that is used does express the values held. (When elements of our natural surroundings do begin to matter to us, the scientific terms in this field also receive their appreciative valuation. Think, for instance, of 'sustainable environment', 'biodiversity' or 'radioactive'). Scientists who try to avoid appreciation or depreciation often find it problematic to find new, more-neutral wording, particularly when their study is on some controversial subject. A nice example is provided by the actions of The Dutch Scientific Council to the Government. In recent decades, it has repeatedly in its advisory reports tried to introduce a new vocabulary to refer to immigrants and their descendants. In the early 1990s, it moved from 'ethnic minorities' to 'allochtones'. Recently, it shifted again to 'individuals having a migration background'.

Furthermore, research findings are often diverse and complex. They contain information on several tested hypotheses or a series of items; information on exceptions and complicated cases; and information on the limitations of a study. Advisors cannot present the entire complicated picture but must select and note 'what is relevant'. From a study into the level of civic participation in pension reforms, they might, for instance, select findings that are of relevance for a policymaker involved in pension policy. However, they might also use some of their findings for a report on interactive policymaking for a completely different field.

The recent shift in research towards using big data makes the reality of this choice even more apparent. The analysis of large amounts of data can produce a wide variety of correlations. Data analysts who want to use their findings for advice must decide how to translate their findings to a particular policy field and turn it into advice. A correlation between social background and formal complaints made against public officials might be interpreted in many different ways and linked to a diversity of policies. (Here, however, it might be argued that epistemological values are not accurately observed. The openness of interpretation is a consequence of an inductive procedure followed in this type of big-data analysis and violates basic rules of falsificationism.)

In all such cases, researchers select from their research findings and thereby choose to be inconvenient for one and not the other policymaker.

Modes of Inconvenience

Although remaining within Weber's ideal of bringing inconvenient truth, scientific advisors retain options on how to employ this truth. Initially, they might provide their empirical knowledge so that it invites the policymaker to adjust the use of a particular policy instrument. They might use their findings on the effectiveness of the use of social media compared with older media in influencing people for advice to a municipality. 'Change your communication strategy on rubbish and waste disposal. If you start informing citizens via social media, you will communicate your message more effectively.'

Researchers can also provide advice so that it questions the use of a certain instrument and suggests the use of another. For instance, they might note that their research shows that the effectiveness of all communicative instruments is quite low and suggest the introduction of a legal instrument. In this case, the change advocated is more fundamental. It involves a shift to another instrument that in fact implies a change in the understanding or valuation of the issue at hand. If the advice is followed, the actions of certain citizens (for instance, not placing their rubbish bins on the pavement) change from actions that are undesirable to actions that are criminal. A similar shift can be noted in cases in which the advice that points to a more efficient instrument entails a shift from public provision to private provision. In quality and availability, the service might remain the same, but its meaning and its valuation changes; a public good becomes a commodity. In these cases, the scientific advisor chooses to present inconvenient facts that in a certain sense imply a change in policy aims.

Such an effect is clearly present in a further mode of providing inconvenient facts. Scientific researchers might note that their findings imply that the policymaker's aim can only be realized at great costs. An investigation into the costs and consequences of a ban on the production and sale of cannabis, for instance, might invite policymakers to re-evaluate their aversion to soft drugs and reconsider their priorities in fighting the problem. Scientific advice might also bring to the policymaker's mind aims and values not considered previously. A study of the effects of introducing some type of electronic highway toll-system might note that it requires gathering a wide variety of personal information and thus infringes upon citizens' privacy. Such a study might induce the policymaker to consider the value of privacy and, again, lead to a re-orientation of priorities and valuations in addressing transport. In general, scientific advice can imply that certain interests that were initially neglected (for instance, of particular minority groups, animals, and future generations) or values that were not considered (e.g., privacy and sustainability) now do become part of the picture.

Not every study contains the sort of findings that make all of these types of advice possible. However, scientists clearly have a choice; they can employ their inconvenient facts differently. They can opt for a mere calibration of instruments and more efficiency, but they might also open the policymaker's eyes to other aims and values than had been considered previously. In the latter case, the scientific advisor invites or seduces policymakers into reflection upon their whole policy frame (compare on this latter possibility Schön and Rein 1994).

Structuring Problems

Weber's image of policy advice pictures a policymaker that has some policy preference in terms of aim and instruments. That representation of reality often proves to be much too simple, as many researchers of the field have noted (e.g., Majone 1989; Vickers 1995; Vining and Weimer 2017). Wildavsky, for instance, maintains that the scientific advisor, in fact, often must define the policy problem at hand. Policymakers themselves often do not have a clear idea of what exactly the problem is when they invite scientific advice on some matter. Policymakers are often confronted with a diversity of demands, values, and constraints in a certain area. There are complaints, there are indications that current policies do not deliver on their promises or that they do not fit changed circumstances, and all types of demands and claims are being made. Consider a municipal policymaker that must address the implementation of a major decentralization in the field of youth care. The municipality already had tasks in this field, but national government hands over tasks, responsibilities, and budgets. All types of observations support the idea that the old way of doing things does not work anymore. When policymakers turn to the advisor for insight, they actually ask the scientist to structure the policy issue (Dunn 2012: Chap. 3).

This structuring implies identifying or specifying the problem—what is wrong in the current situation given which particular value or aim? However, it also means identifying a problem that can be the object of a policy. Some phenomena might not suit us. Nevertheless, in a given setting, we might not be in a position to change anything about them. Policymaking, then, is not an issue. Structuring a policy problem means formulating the problem so that it can have a solution, a puzzle that can be solved (Wildavsky 1987: 3, 5). It is to '[c]reate problems that decision makers are able to handle with the variables under their control and in the time available' (idem p 15). Dunn notes that the advisor's task in such policy-structuring cases is to present 'the most appropriate formulation of the problem' (Dunn 2012: 73).

In the formulations of both Wildavsky and Dunn, scientific advisors clearly must make a choice or to perform a judgment. They must choose, from the different possibilities they can develop, the problem formulation that can be an option for this policymaker (Wildavsky); in other words, they must judge which one is most appropriate (Dunn). Whatever this approach exactly implies, it is not a matter of employing some epistemological value or method.

Some authors in the field of policy analysis suggest that aiding policymakers in elaborating and developing their understanding of policy problems can be the only task of external experts. They are quite sceptical about the possibility of any scientific knowledge being actually relevant to policymaking. The role of experts, they maintain, is limited to helping policymakers to make explicit the tacit knowledge they already have and help them structure problem-frames (Lynn 1996; Schön and Rein 1994: 196). In this book, the issue of whether, and to what extent, scientific knowledge can be of help to policymakers will not be investigated. It will simply be assumed that scientific knowledge, of any type, can be of use for policymaking.

It may be concluded that translating research findings into policy advice means making certain decisions. For making these decisions, scientific advisors cannot rely only on epistemological values and research methods. Such decisions involve more than being honest with the facts; they imply choosing how to contribute to what policy issue or social problem in what way. Furthermore, if we follow the observation of some scholars in this field, that is, that policy processes are often not linear processes and that policy advisors are involved at multiple points in them, the number of choices for the advisors only multiplies (e.g., Bijker et al. 2009).

3 Simply Bringing Truth to Power?

There are further choices that scientists as advisors must make. They must also decide how much effort they will make to bring their inconvenient facts into the policy-making domain. They might wait until they are invited to present their expertise and, then, do so without making much of an effort to convince policymakers. They can also choose to be more active in the presentation of their message and to take greater care of convincing their public. A greater concern for the rhetorical effect of their message means that scientists should not only mind that their advice is scientifically sound and logical in argument. Apart from this aspect of *logos*, as Aristotle called it, good rhetoric also involves two further ingredients. There is *ethos*: ensuring that one can speak with authority, that one is considered trustworthy. A further element is *pathos*, which involves touching emotional strings that help to convey a message (Aristotle 1991). In the literature, many types of reasons are mentioned why advisors should employ rhetorical skill to be effective. Some note that power holders might be inclined to understand expert advice, particularly from outsiders, as a threat to their position. Part of the power holder's legitimacy might lie in their (alleged) expertise, which might appear to be undermined by taking advice, as Machiavelli suggests (Tholen 2016). Others emphasize that policymaking is not only about rational deliberation but also about politics. Those who challenge a policy position can be viewed as (potential) political enemies (Dobel 1999). Others still show that policymakers tend to view certain policy aims as inseparably linked to certain policy instruments (Majone 1989). Economic growth might be simply identified with lowering taxes; lower crime with severe punishment; good education with state control of schools. These policymakers are inclined to dismiss inconvenient truths about the effects of those instruments as comments on policy goals. As a last example from the literature, there are analysts who note how organizational routines and groupthink create barriers to actually hearing inconvenient messages (March and Olsen 2006; Janis 1972).

Scientific advisors must choose whether, and to what extent, they will employ rhetorical tools to communicate inconvenient truths to the mind of the policymaker. They might decide to use all types of communicative techniques. They can also choose to translate their advice into a frame with which the advised will or will not feel comfortable.

A further type of choice scientific advisors must make, in terms of activism, concerns their choice of venue. They can place inconvenient truths before a lower-level policymaker. They might, however, also present their findings to a higher-level politician (or threaten to do so). They can go even further and take their results to some controlling body, such as parliament. To have their material taken seriously, they can provide it to a member of the majority parties but also to the opposition. In addition, there is of course the option of going to the media. The point here is not that scientists should always take these highly activist roles in delivering inconvenient truths. Nevertheless, it is an option that they can consider. It is an option that leaves their epistemological values and intellectual honesty (as Weber posits) intact.

4 Summary

This chapter addressed the choices a scientific advisor must make in the last phase of the process. Here again, choices involve considerations in which non-epistemological values must play a role. Epistemology and research method are of little help in determining what selection to make in one's findings, to which policy to relate the findings, and what level of activism to employ in communicating the inconvenient truth to the attention of power.

References

Aristotle. (1991). *The art of rethoric*. London: Penguin Books.

Bijker, W., Bal, R., & Hendriks, R. (2009). *The paradox of scientific authority. The role of scientific advice in democracies*. Cambridge, MA: MIT Press.

Dobel, J. P. (1999). *Public integrity*. Baltimore: Johns Hopkins UP.

Dunn, W. N. (2012). *Public policy analysis*. London: Routledge.

Dupré, J. (2007). Fact and value. In H. Kincaid, J. Dupré, & A. Wylie (Eds.), *Value-free science? Ideas and illusions* (pp. 27–41). New York: Oxford University Press.

Janis, I. L. (1972). *Victims of groupthink: A psychological study of foreign-policy decisions and fiascoes*. Boston: Houghton.

Lynn, L. E. (1996). *Public management as art, science and profession*. Chatham, NJ: Chatham House Publishers.

Majone, G. (1989) *Evidence, argument, and persuasion in the policy process*. New Haven: Yale University Press.

March, J. G., & Olsen, J. P. (2006). The logic of appropriateness. In M. Moran, M. Rein, & R. E. Goodin (Eds.), *The Oxford handbook of public policy* (pp. 689–708). Oxford: Oxford University Press.

Schön, D. A., & Rein, M. (1994). *Frame reflection. Towards the resolution of intractable policy controversies*. New York: Basic Books.

Tholen, B. (2016). Machiavelli's lessons for public administration. *Administrative Theory & Praxis, 38*(2), 101–114.

Vickers, G. (1995). *The art of judgment. A study of policy making*. Thousand Oaks, CA: Sage.

Vining, A. R., & Weimer, D. (2017). Policy Analysis: A valuable skill for public administrators. In J. C. N. Raadschelders & R. J. Stillman II (Eds.),

Foundations of public administration (pp. 162–176). Irvine, CA: Melvin & Leigh.

Wildavsky, A. (1987). Analysis as art. In A. Wildavsky (Ed.), *Speaking truth to power. The art and craft of policy analysis.* New Brunswick, NJ: Transaction Publishers.

CHAPTER 6

Addressing Non-Epistemological Issues

Abstract From the previous three chapters, it follows that in scientific advice, many decisions must be made for which scientific advisors cannot exclusively rely on epistemological values. For two reasons, this requirement might be a problem. First, it appears to be at odds with democracy; decisions with important societal implications are made by a select group. Second, such decisions often appear not to be made deliberatively and, therefore, possibly irresponsibly.

Keywords Responsibility · Democracy · Social values

1 Value-Free Science?

In has now become clear that speaking about value-freedom in science can be misleading. First, science cannot exist without values that guide its methods and logic. Only because scientists follow epistemological guidelines and honour the values that they express can they distinguish better from worse statements about how things in the observable world are and how they came to be that way. There is, to be sure, an ongoing debate over what exactly these values and methods encompass. Some focus on rules that lead to better theories and general causal explanations. Others present guidelines that help to obtain ever-better interpretations of what phenomena in the world mean for particular people in particular contexts. However, such different approaches agree that there

B. Tholen, *Virtue and Responsibility in Policy Research and Advice*,
DOI 10.1007/978-3-319-65253-5_6

are epistemological values that can be and must be distinguished from social values. The latter values do not address proper knowledge but rather address for example what we value in the world, what is vulnerable and deserves our attention and concern, and how a more just or free society would appear.

These social values, and that is the second misleading aspect of the term value-free science, also have their relevance for doing research and for scientific advice. Scientists and advisors must make all types of decisions: choosing the research topic, formulating the research question, selecting relevant factors and theory, choosing an instrument for measurement, considering what level of risk of being wrong to accept, deciding what inconvenient facts to bring to whom and in what way. For all such questions, epistemological values alone do not offer sufficient guidance. Considerations concerning the possible effects on what is valuable, that is, social values, come into play.

A value-free science can only be a science full of values. However, the concept of a value-free science can remain a meaningful concept if it is understood in a more sophisticated way; that is, in fact, how Weber used the phrase. Value-free science is science in which values have their proper place. Epistemological values guide the process of observation, testing theories, and developing better interpretations. Deviating from these guidelines, for instance to obtain quick or favourable results, means fraud, cheating, and misleading. Nor might these decisions be influenced by social values. Anyone who rejects a theory, not because it is at odds with systematic observations, but because it does not fit one's ideas about how the world should be, misunderstands the epistemological core of science. This point is what Weber's intellectual honesty encompasses. Social values and politics, here, are off limits.

However, what about these other decisions, the ones concerning the knowledge that the advisor eventually can produce that demand the employment of social values? Do they not make scientific advice very socially or politically value-laden? Here, two issues must be addressed. First, sociologists and historians of science suggest that the decisions noted in the last three chapters are often routinely made. Individual scientists follow their colleagues, employ routine practices, or express tacit knowledge (e.g., Jasanoff 1990; Kuhn 1996; Latour 1987). In research reports, this type of choice is hardly mentioned, let alone explicitly reflected on (e.g., Douglas 2009). If this point is indeed true, it invites the question, is it not problematic that scientists do make such decisions, possibly with large policy consequences, without much thought? Is not such behaviour acting irresponsibly? Second, the overview of decisions that cannot be guided by epistemological values but still influence policymaking appears to imply that scientific policy advisors are in fact policy advocates. Their decisions, whether or not made reflectively, determine what (more or less) inconvenient truth on which policy issue will be helping, or hindering, which political interest. The scientists' choices set the political agenda, steer political debate and influence decision-making. Their value-choices determine, to some extent, politics. Is that not undemocratic?

These issues will be further elaborated in the next two sections.

2 Science as Politics?

According to a variety of scholars, the impact that the choices of scientists can have poses a problem. The non-scientific decisions—for example, on research topic, risk, and activism—draw science into the field of politics. The choices prioritize certain values and not others, and they come to the aid of some interests and set others back. Two sorts of concerns can be distinguished on this point. A first concern is that scientists might obtain a disproportionate voice in democratic policymaking. Scientists have the policymaker's ear. The choices scientists make and the advice they thus can provide have an impact that far exceeds the opportunities of many other societal groups. This unequal input, this privileged opportunity structure, results in democratic unfairness (e.g., Brown 2009; Kitchener 2001: ch 11). Moreover, the scientists' influence entails a bias in policy and politics. Some studies suggest that scientists differ in their political orientation from the average citizen. Scientists in certain areas are said to be left-wing voters more often than are the public overall.

A different comment of bias focusses on the basic orientation most scientists have towards their role and function. For instance, such an analysis can be found in the work of Habermas. Commenting on Weber's approach, Habermas maintains that a science that considers itself value-free in fact expresses a particular value or interest. The science that Weber promotes provides information that can be translated into technical recommendations for a choice of means to a particular end.

It produces knowledge that can be used technically, and it expresses a technical cognitive interest (Habermas 1971). In other words, although it presents itself as neutral, it is in fact biased towards intervention and effective control. A somewhat similar diagnosis is presented by Foucault. Foucault also notes how the sciences, particularly the social sciences, have provided the type of knowledge that enabled social and political institutions to structure and control our lives. In fact, they have provided knowledge and techniques, leading us to further self-control that disciplined us (Foucault 1975). In Foucault's work, a further comment on scientific bias can be found. It is a comment that is shared by feminist and cultural studies. In mainstream science, this critique maintains, scientists in their non-epistemological decisions all follow some dominant pattern. They are inclined to choose for example research topics, categorizations, and labels that, in fact, express and support the status quo. Existing gender relationships and racial relationships are not challenged but rather are tacitly acknowledged. Consequences of policies, particularly for women and ethnic minorities, subtle and unconscious approaches to selection, and particular viewpoints are in fact excluded in common formulations; such issues are generally neglected in scientific research (Foucault 1975; Longino 1995).

Different as these comments might be, they all imply that scientific knowledge always reflects a certain interest or contains a systematic bias. They all opt for making choices in science that remedy the bias—by shifting to an emancipatory cognitive interest in science (Habermas) and by moving to a critical science that reveals dominating patterns and disciplining discourses in society (Longino).

Some of the claims these critics make might be questioned. Why should the Weberian practice of presenting inconvenient truth imply that it expresses technical interest? What about scientific findings that in fact have led to a disruption of the status quo and dominant power relationships (such as the invention of birth control pills)? Should all research really be critical and emancipatory; is effective control not also of value in many circumstances? Such questions, however, do not touch the main point that the critical studies entail—that the non-epistemic choices that scientists must make might have political implications. What, then, would be the proper way to address this type of choice? However, before we can turn to this question, another issue requires further elaboration.

3 (Ir)Responsibility?

All concerns about the non-epistemic decisions that must be made are related to the consequences of the choices scientists make. The following question might be posed: why should a scientist be worried at all about the consequences of his findings and the presentation thereof? One might imagine a scientist saying, I merely provide knowledge. I do not know what others are going to do with it or what its impact is going to be. I can be held in no way responsible for the consequences, so why should I bother? One might be inclined to retort to such a scientist, why present your findings, or even, why do research if it is not for changing the world for the better? If something is worth knowing, is it not then also worth preserving? There is, however, also a more substantial answer possible.

Scholars have answered the imaginary scientist by elaborating on the idea of responsibility of scientists. Thompson and Douglas, for instance, vest the moral responsibility of researchers on the same ground as the responsibility we attribute to all people. We commonly consider free and reasonable individuals answerable for their actions and inactions. They maintain that there is no reason to exclude scientists and scientific advisors from this class (Thompson 1983; Douglas 2009). In some areas of research, to be sure, it might be granted that studies hardly have any direct consequences or imaginable risks for human beings and their environment. This statement might be true, for instance, in the fields of cosmology and mathematics. In other branches, however, and particularly in fields that offer policy advice, the possibility of having an impact is very real. Of course, scientists might argue that their choices have only little effect on the policy decisions that are being made. They are not the ones who have the authority to make the policy decisions; decision-makers must make up their own minds. Many others in the policy process, moreover, will also provide their input. Scientists might acknowledge their responsibility in the treatment of test-animals or in the role people obtain in experiments, but maintain that they cannot simply be held accountable for anything that others do with their findings. In this reply scientists have a point. Their responsibility depends upon their actual causal role and abilities and on their liberty to act. The researcher is not the policymaker, and research conclusions, or even policy advice, may not be equated with policies or policy outcomes. However, that still does not mean that scientists have no responsibility for what their studies entail. Thompson notes two types of cases in which it is evident that the researcher can be held ethically accountable. First, he sketches the case of scientists who, because of their particular expertise, are the only ones who are aware of some present danger or possible harm. Few would doubt that they should be held accountable if they did not act upon their knowledge and warn those involved. A second typical case is that of policymakers who choose a disastrous line of policy, an approach that they would not have taken were it not suggested by the scientific advisor. In such a case, the advice of this scientist was a sine qua non, a necessary link in the causal chain. Here, again, is a strong case for holding the scientist accountable (Thompson 1983). These two cases emphasize the general idea that the scientist and even more so the scientist advisors are responsible in principle for their acts qua scientific advisors. Having no formal decision-making power does not exclude them from responsibility. Of course, in cases other than the clear-cut ones that Thompson provides, the level of responsibility of the advisor might be smaller, and it might be much more difficult to determine. However, the conclusion is that even when we follow the minimal understanding (responsibility as mere accountability), the responsibility of scientific advisors concerning the social consequences of their research and advice cannot be dismissed.

4 What Is Next?

The question that now must be answered is, how should a scientific advisor address the issues that epistemological values (alone) cannot answer? What is an appropriate way to address the issues that imply inclusion of social values? In the next four chapters, four types of answer will be investigated: a scientific answer (Chap. 7), a normative answer that offers democracy as a solution (Chap. 8), and two further normative answers that find their inspiration in ethical traditions (Chaps. 9 and 10).

References

Brown, M. B. (2009). *Science in democracy. Expertise, institutions and representation.* Cambridge, MA: MIT Press.

Douglas, H. E. (2009). *Science, policy, and the value-free Ideal.* Pittsburgh, PA: University of Pittsburgh Press.

Foucault, M. (1975). *Discipline and punish: The birth of the prison.* New York: Random House.

Habermas, J. (1971). *Knowledge and human interest*. Boston: Beacon Press.

Kitcher, Ph. (2001). *Science, truth and democracy*. Oxford: Oxford University Press.

Jasanoff, S. (1990). *The Fifth Branch: Science Advisers as Policymakers*. Harvard University Press (Cambridge, Ma.).

Kuhn, Th. (1996). *The structure of scientific revolutions*. Chicago: University of Chicago Press.

Latour, B. (1987). *Science in action*. Cambridge, MA: Harvard University Press.

Longino, H. E. (1995). Gender, politics and the theoretical virtues. *Synthese, 104*, 383–397.

Thompson, D. (1983). Ascribing responsibility to advisors in government. *Ethics, 93*(April), 546–560.

CHAPTER 7

Solving the Issue by Distinguishing Types of Problems

Abstract A first answer to the issues noted in Chap. 6 can be found in an approach that is common in the Public Administration and Policy Analysis literature. It maintains that the type of problem must be guiding for addressing the value choices. On closer inspection, this approach proves to be problematic. At issue are the objective definition of problems and the justification of guidelines.

Keywords Problem approach · Advisory role · Value free

1 Introduction

In this chapter, a particular strategy to address the decisions that were identified in the previous chapters is investigated. This strategy is regularly presented, or implied, in the Public Administration and Policy Analysis literature. This approach gives central importance to the type of problems that scientific advisors must address. It maintains that particular types of problems should be addressed, each in a specific way. In the first part of this chapter the general characteristics of this approach will be sketched. Next, the approach will be assessed critically by examining more closely the two central elements, that is, the definition of problems and how the problem types are used to instruct policy advisors.

B. Tholen, *Virtue and Responsibility in Policy Research and Advice*,
DOI 10.1007/978-3-319-65253-5_7

2 Problem Types and Advisory Roles

In the Public Administration literature, a variety of approaches can be found that express a similar claim; characteristics of policy problems determine the proper answers to questions that scientific advisors face (e.g., Bijker et al. 2009; Douglas and Wildavsky 1983; Dunn 1988; Ezrahi 1980; Hoppe 2010; Pielke 2007). Those questions, in fact, belong to the set of non-epistemological issues that were identified in the previous chapters. They mostly involve the level of scientific activism, the involvement of scientists in reflection on policy aims, and favourable types of research design.

A core element of the Problem approach is the distinction between different types of policy problems, which often involves a categorization of levels of consensus on problems. The work of Hoppe provides a good example. Hoppe distinguishes four types of problems—a typology that is based on the combination of two dimensions. One dimension concerns the level of agreement that exists on the norms and values involved in the policy issue; how much agreement is there on aims? The second dimension concerns the level of certainty on the relevant knowledge; about what are we certain concerning what occurs and how we can intervene? (Hoppe 2010).

The next element in this strategy is linking such a categorization of problem types to particular roles or tasks for scientist advisors. Pielke, for instance, distinguishes between the Pure Scientist, the Science Arbiter, the Issue Advocate and the Honest Knowledge Broker. Each of these roles fits a particular problem type best. If there is great disagreement on values, for instance on an abortion policy, the advisors should not act as a mere presenter of technical information but become involved as an Issue Advocate. Moreover, if there is agreement on the policy aim, the scientist should not act as an Honest Broker who expands the scope of choice for the decision-maker (Pielke 2007: 1–3). Hoppe makes a similar point when he maintains that a Rational style of advice is appropriate for problems with a high level of consensus. Employing this style means performing cost-benefit analyses, audits, and programme evaluations, that is, providing evidence-based policy input. However, if there is only a moderate level of consensus on aims, then Client advice or Process management styles are advisable. Characteristic of both styles is providing strategic advice that helps policymakers to realize their aims. Different strategies fit the other problem types. A good advisor recognizes problem types and chooses the proper style of advice accordingly (Hoppe 2010: 116, 188–194). Conversely, bad advisors neglect the problem type in their choices and face wrong problems (Dunn 1988). Their advice is counterproductive for public debate and political choice (Hoppe 2010: 6).

3 Critical Questions

For a critical assessment of the Problem approach, two issues must be further investigated. One issue concerns the categorization of problems, the other the linking of problems to roles of policymakers and styles of advice.

On Categorizing Problems

Prima facie, it appears an attractive and reasonable recommendation—'one should do what the circumstances demand'. However, on closer inspection, it becomes less clear what this advice entails in a particular policymaking environment. What does it mean to have consensus on a policy issue? The studies in Public Administration mentioned above are not clear and even contradict one another. For instance, does consensus on aims mean that people agree on what particular aim should be given priority or that they agree on what values are relevant for this issue? Does uncertainty on the knowledge-dimension refer to the lack of any (reliable) information, to an ambiguity of the 'facts', or to the level of risk of failure that can be attributed to the employment of certain instruments?

The strategy becomes even less clear when one focusses on the setting or scale of a problem. Should one be considering consensus and disagreement in political arenas or in a broader societal context? Some issue might be hotly debated in parliament, whereas it is not really considered a problem in society at large. Moreover, particular social concerns might never find their way to the political arena. Gusfield's analysis of the different ways drunk driving can be understood as a problem and of what problem definitions did reach the political agenda offers a nice illustration on this point (Gusfield 1981). The Problem approach does not answer the question, whose understandings should be guiding in determining the character of a problem. The consensus in exactly which arena, area, or domain is relevant?

The vagueness in the concept of consensus is not the only complication that haunts the Problem-type approach. A further issue becomes apparent when one realizes that (public) understandings of issues might be subject to manipulation. Particular actors can have an interest in trying to influence the (apparent) level of consensus. Politicians might want to evade or neglect socially controversial issues that they all think dangerous for their electoral position. Tobacco companies might support dissenting voices to break a developing scientific consensus on the harm of smoking (Ezhari 1980). All types of actors might be motivated to employ means to change the level of consensus, or at least the public understanding of the level of agreement, among, for instance, scientists. Thus, determining the real or true level of consensus on some issue itself becomes a problem.

Underneath this latter point, there lies a more fundamental issue—determining whether and when there is a social or policy problem (not to complicate the issue further, I will simply equate social and policy problems here). The literature on Problem approaches is relatively short on explaining what a policy problem is. It is merely implied that in the case of such a problem, there is some clear common understanding that some state of affairs is not desirable and that some intervention might be appropriate. It remains, of course, quite vague what 'common' here means. Furthermore and more fundamentally, it is unclear why there should be an explicitly shared and specified notion of what is undesirable or problematic. In a particular context and time, heavy drinking, child beating or wife battering might occur without being understood as social problems in the above sense. However, scientists might study them, present their results, and thereby arouse a common sense that there is a (hidden) social problem (Root 2007). Thus, it actually becomes the choice of scientists whether the relevance and importance of an issue as 'a problem' is to be defined by some public acclamation or by their own judgment in the light of societal values.

A basic problem underlying the Problem approach, therefore, is that it must presuppose (the character of) policy problems as a given for policy advice, whereas such advice at the same time can change the character of a policy problem.

On Linking Problems to Advisory Roles

The second part of the Problem approach also has its complications. A first issue concerns the actual empirical support that advocates of a Problem approach can provide for their claims. The authors mentioned above actually offer little empirical evidence (see Tholen 2016). The claims that are made, moreover, are not evident. Why should one, for instance, act as an Honest Broker and try to broaden the issue when there is great normative disagreement, as Pielke posits? In some cases, the public disagreement on some issue might disappear when some authoritative body presents an answer. It might also be the result of presenting new scientific findings (by the Pure Scientist, the advisor that uses the Rational style) that people change their minds on issues on which values differ. Even on deeply contested issues, such as the abortion example that Pielke discusses, new medical findings and techniques might make people alter their views (compare Brown 2008; Jasanoff 2008). Disentangling issues or cutting problems into elements that can be addressed separately might be satisfactory. Wildavsky's words, mentioned earlier, might be quoted here: problems must be created that can be solved (Wildavsky 1987: 17). Palonen, to mention a last example, shows how particular problems might be approached in a 'Machiavellian way' (creating order in a setting that is generally understood as chaotic and unpredictable) or a 'Weberian way' (cutting through routine understandings and predictability). In such approaches, the problems are not taken as a given but, in a sense, created in the advisory strategy (Palonen 1998).

The last examples imply a further question. The Problem approach demands that scientific advisors respect the existing (public) consensus on an issue. However, why should they do so? Why would it (always) be wrong to choose a style that disrupts the consensus on some issue? Researchers might have found that a particular policy generally has beneficial consequences, but the negative implications for a particular group (e.g., migrants and women) have been neglected thus far. Their findings might imply that particular practices or states of affairs encompass great risks and possible harm that have not yet been recognized in political or public arenas. Moreover, it might not only be specific consequences that scientist feel are neglected; they might also witness tunnel vision in discussions and problem definitions. For instance, Burawoy emphasizes in his sociological address that a concern for state and market dominates public discourse at the expense of civil society (Burawoy 2005). The advocates of the Problem approach do not offer an argument for the demand to respect the existing consensus. The implication of this demand, however, is that this approach has a strong conservative (or conventionalist) bias.

This observation leads to the insight that the Problem approach contains or is in fact built on a particular value-orientation. The basic idea is that explicit and common ideas (among policymakers, or the public at large?) should be guiding for the choices of scientific advisors. If there is consensus, they should not interfere. If there is no commonly felt urgency about some harm or wrong scientists think real, they should not intervene. Good arguments might be given against such a stance, but that is not the issue here. The point is that the Problem approach expresses a particular value-choice and is not, and cannot be, the mere conclusion of an empirical or logical scientific analysis. (Hoppe, to be sure, shows himself to be aware of the ethical base of his approach (Hoppe 2010: 7–8). However, it remains unclear how his particular understanding of (ethical) human nature can support that approach.)

4 Conclusion

The Problem-type approach that is presented in the Public Administration and Policy Analysis literature suggests a convenient strategy to address some of the choices that scientific advisors must make. The conclusion must be that this approach does not and cannot lift the task of making value-choices from the shoulders of the scientists. First, if one decides to follow this strategy and determines that problem types and advisory roles are strongly connected, one still must face some value-issues. One must answer the question of who should be included in the forum of consensus/disagreement. This answer will express one's understanding of relevant interests and the appropriate arena. Furthermore, one must decide what counts as a real consensus and what as an improper one, created by special interests. Here, again, a particular value-orientation must be guiding—in this case on fair interest representation.

Second and more fundamentally, the choice for the problem strategy itself is also a value-choice. Advocates of this approach suggest that the choice is of a scientific type. Scientific facts simply support it, their work appears to imply. To be sure, the approach does contain an empirical claim: 'taking these advisory roles has those effects'. Basic to this claim, however, is a particular understanding about what a social or policy problem is and (thus) what it means to solve it. A social problem appears to be an issue that at some point is generally recognized as an undesirable state of affairs, and that problem definition itself must be understood as a given. That this problem definition, in fact, is a value-choice becomes clear when one considers that problems might also be understood as not (yet) recognized and that current problem definitions need not be taken as given. In other words, the Problem approach has a particular (conservative) value-core.

The Problem approach suggests that answers to non-epistemic choices can be deduced from scientific findings. However, this example illustrates that such a shortcut cannot be provided.

References

Bijker, W., Bal, R., & Hendriks, R. (2009). *The paradox of scientific authority. The role of scientific advice in democracies.* Cambridge, MA: MIT Press.

Brown, M. B. (2008). Review of R.A. Pielke; The honest broker. *Minerva, 46,* 485–489.

Burawoy, M. (2005). *For Public Sociology. American Sociological Review, 70*(1), 4–28.

Douglas, M. & A. Wildavsky. (1983). *Risk and culture.* Berkeley, CA: University of California Press.

Dunn, W. N. (1988). Methods of the second type: Coping with the wilderness of conventional policy analysis. *Policy Studies Review, 7,* 720–737.

Ezrahi, Y. (1980). Utopian and pragmatic rationalism: The political context of scientific advice. *Minerva, 18,* 111–131.

Gusfield, J. R. (1981). *The culture of public problems: Drinking-driving and the symbolic order.* Chicago: University of Chicago Press.

Hoppe, R. (2010). *The governance of problems. Puzzling, powering and participation.* Portland: The Policy Press.

Jasanoff, S. (2008). Speaking honestly to power. Review of: The honest broker: Making sense of science in policy and politics, by R.A. Pielke. *American Scientist, 96*(3), 240–243.

Palonen, K. (1998). *Das 'Webersche Moment': zur Kontingenz des Politischen.* Opladen: Westdeutscher Verlag.

Pielke, R. (2007). *The honest broker. Making sense of science in policy and politics.* Cambridge: Cambridge University Press.

Root, M. (2007). Social problems. In H. Kincaid, J. Dupré, & A. Wylie (Eds.), *Value-free science? Ideals and illusions* (pp. 42–57). Oxford: Oxford University Press.

Tholen, B. (2016). The value of the issue context approach for scientific policy advice. *Science and Public Policy, 43*(2), 184–191.

Wildavsky, A. (1987). Analysis as Art. In: A. Wildavsky (Ed.), *Speaking Truth to Power. The Art and Craft of Policy Analysis.* New Brunswick, N.J: Transaction Publishers.

CHAPTER 8

Solving the Issue by Bringing in Democracy

Abstract A second answer to the problems summarized in Chap. 6 involves deciding democratically on value issues in science and scientific advice. Such a democratic solution might take many forms and follow different ideals of representation. However, no democratic arrangement will be able to remove all of the decision-making tasks from the shoulders of scientific advisors. In fact, democratization creates new non-epistemological choices for scientists, for instance on tutoring laypersons on scientific choices.

Keywords Democracy · Representation · Tutoring

1 Introduction

The last chapter focussed on an approach that essentially maintained that one can turn to scientific findings to resolve the issues that must be answered in the course of research and advice. Such an approach proved questionable; value-decisions cannot be solved or even evaded with that approach. The approach that is investigated in this chapter adopts the opposite course. It maintains that the non-epistemological decisions in science and advice should not be in the hands of scientists but must be placed before a democratic forum. A basic ideal of our democratic society is that decisions that have serious consequences for society should be

B. Tholen, *Virtue and Responsibility in Policy Research and Advice*,
DOI 10.1007/978-3-319-65253-5_8

made so that the values and interests of all people are considered. Why should that be different for these scientific decisions?

Within the study of Public Administration and Policy Analysis, such concerns—in combination with others—have motivated scholars to involve people actually affected by the process of research and advice. Particularly in the field of environmental planning and neighbourhood development, researchers have, to a greater or lesser extent, given laypersons a role (e.g., Fischer 1993; Flyvbjerg 2001; Jennings 1987; McSwite 1997; Wagenaar 2011; compare Tholen 2015). A truly democratized science, however, would affect all non-epistemic decisions in every study.

To understand and evaluate this type of answer to the problem, I will concentrate in this chapter on the well-elaborated argument for democratized science by Philip Kitcher (2001, 2011). In his recent books, Kitcher has recognized the variety of value-decisions that must be made in research and advice. He argues that democracy in science should be of a particular type. If it is not, it simply reproduces a type of decision-making that he calls elitist. In the first section of this chapter, Kitcher's ideal of a well-ordered science will be presented. Next, this position will be assessed.

2 Advocacy for Democratic Involvement in Scientific Advice

In his *Science, Truth*, and Democracy, Kitcher (2001) elaborates on what it means for scientists to decide what they should study, that is, to determine what counts as significant knowledge. In *Science in a Democratic Society* (2011), he adds further questions to the list the scientists must address: what level of risk of being wrong is acceptable, and how should the results be brought to the public and applied. Kitcher thus touches upon the decisions in all of the three phases that were identified in the earlier chapters of this book. Many scientists and others misunderstand, he claims, the character of such decisions. They interpret them to be issues that can be decided upon by employing scientific findings and scientific method and thereby suggest that the agenda for science must be and can be objective or neutral. This scientism, however, is mistaken. It neglects that scientific knowledge and method alone are not capable of determining what knowledge is significant, when it is good enough, and how it should be applied and why. It is blind to the role of social values in scientific decisions.

The decisions that are made in science can have serious consequences for groups of people or even for humanity at large. Kitcher notes how the needs of disadvantaged groups are often neglected by research agendas. Medical research, for example, is more focussed on the problems of relatively few elderly people in rich countries than on cures for diseases that are broadly dispersed in poor countries. He also emphasizes how scientists are often pressed into fostering unrealistic expectations of practical benefits to secure public funding (2001: 127 ff).

A proper approach to making scientific decisions starts from the insight that these decisions have social and political consequences. In modern society, these decisions, like other decisions with such an impact, must be made democratically, Kitcher maintains. However, leaving all decisions simply to the people is not what he proposes. Such a 'vulgar democracy' should be avoided, Kitcher argues. A simple system of voting by a largely uninformed public would mean a 'tyranny of the ignorant'. Furthermore, a proper democratic system should ensure that there is no situation in which only a tiny fraction of the population in fact decides. Democratic decisions, third, should be based on real consideration of alternatives and not on fixed frames of problems and solutions (Kitcher 2001: 126 ff). Kitcher sketches the ideal of an 'enlightened democracy' that encompasses decision-making based on deliberation by representative, well-informed individuals. His argument resembles the one that John Rawls presented in *A Theory of Justice* (1972). Kitcher calls his ideal 'well-ordered science', which is reminiscent of Rawls' 'well-ordered society', and he asks his readers to imagine an ideal setting for deliberation and decision-making. Science is well-ordered, he maintains, 'when its specification of the problems to be pursued would be endorsed by an ideal conversation, embodying all human points of view, under conditions of mutual engagement' (Kitcher 2011: 107).

More particularly, Kitcher asks his readers to imagine a group of ideal deliberators who have widely varying science policy preferences. These participants are tutored initially by scientist experts about the significance of potential studies—that is, the (epistemological) significance for existing research programmes and the possible social value of these studies. The deliberators transform their original preferences into tutored personal preferences, discuss these preferences, and compose a list of desired outcomes. They might come to agree on a research agenda. If they cannot reach consensus, a decision will be made by majority vote. The same type of tutored deliberation and decision-making must occur at an acceptable level of risk of being wrong, that is, concerning what standards of precision and accuracy are appropriate for the particular studies that are chosen. Finally, a third round of tutoring and deliberation and decision-making will be dedicated to matters of information and application of the research findings (Kitcher 2001: Chap. 10; 2011: Chap. 5).

This sketch of well-ordered science is an ideal, Kitcher emphasizes. He hopes, however, that understanding the ideal situation can 'help us to improve our practice' (Kitcher 2011: 116). He does not note how exactly this ideal should be transposed to reality or how practice should appear. Nevertheless, Kitcher's suggestion clearly is that decision-making on value issues in science is better if it comes closer to the ideal situation of a well-ordered science.

3 Complications

Kitcher's ideal of a well-ordered science provides a democratic answer to the issue of value-decisions in science. The answer involves an ambitious idea of democracy that combines inclusiveness and well-informed, open deliberation. The question is, can Kitcher's well-ordered science offer what it promises? Can the non-epistemic decisions indeed be taken out of the hands of scientists, can the downsides of vulgar democracy be avoided, and can this all be done without barriers to scientific advice for policymaking? In this section, a set of complications will be identified.

Tutoring and Selection

Leaving non-epistemic decisions to non-researchers entails the need to tutor these outsiders on many issues. In the first round of deliberation and decision-making, the participants must be instructed on what type of questions science can answer and what sort of knowledge is unavailable for an empirical science. Experts also must show what knowledge is available already, which programmes have proved failures, and what studies might be feasible given the state of the art. The need for this type of instruction was nicely illustrated in the opportunity that the Dutch Academy of Sciences together with Dutch universities offered in 2014 to all citizens to present questions for scientific research. Many questions that were submitted proved to be either scientifically unanswerable or already answered in existing studies.

In the second round of deliberation, a similar picture is found. To decide on the level of 'true enough' knowledge, the deliberators must consider and understand all of the relevant decisions that are made in the course of a particular study. Furthermore, they must understand the methodological background of the choices they make. (For related implications, for instance, of accepting statistical risk, see Douglas 2004: 239). Explaining all of these issues fully in all of their complexity to the deliberators would mean educating these participators to the level of experts in these fields. Tutoring, therefore, will inevitably mean informing the deliberators selectively. This selection, however, can only be made by experts. In any realistic setup, scientists must decide what to present to the lay participants on all of the relevant aspects and how they present it. These decisions, again, are not guided by epistemology and scientific method. Democratizing science means that some value-decisions will be left to (part of) the public, but it entails the introduction of other value-decisions for scientists.

Representation

It is not an option, as Kitcher himself acknowledges, to involve all people. It is not practical to involve all people. Moreover, it is impossible to give future generations a voice—although these are the people who might be most affected by contemporary research. Making (some of the) value-decisions for scientific advice democratically will mean choosing some form of representation. Here, we immediately encounter issues that are central to political theory debate. What type of representation is appropriate; should the deliberative assembly mirror the population in (some of) its characteristics, or should its members merely stand for the interests and preferences of particular groups in society? Should involvement be limited to parties whose interests are directly affected by the proposed research, or should a voice also be given to others? Should the representatives act as trustees who vote according to their own consciences or as delegates who act upon instructions? Kitcher does not begin to answer these questions, and there is no reason to think that the issue of representation is not as difficult to solve here as it is in the general political theory debate. Many different approaches to organizing representation can be imagined (citizen juries, parliamentary committees, expert NGOs, think tanks and boundary organizations, or even a body of scientist peers), but none can be heralded as the best or the most representative (see also Brown 2004, 2013). One might even argue that, given the complexity of tutoring, a representative body consisting only of scientists would have its advantages.

Kitcher's ideal of a well-ordered science is inspired by Rawls' heuristic device of an Original Position. That device is designed to imagine how ideal persons would decide, persons that have all of the relevant knowledge but are unaffected by considerations of power and who are motivated to consider other people's points of view. Rawls does not present his imaginary deliberative session as an ideal for real decision-making bodies. He uses it to present and defend specific outcomes, that is, a set of basic principles for the well-ordered society. Kitcher's argument for a well-ordered science might be interpreted similarly. Such a heuristic device-interpretation would have the advantage that the problem of designing a representative body can be avoided. Scientists who encounter a value-issue in the course of their research should consider what an ideal council would decide. Nor does Kitcher explicitly elaborate this possibility of employing his ideal. However, he emphasizes at one point that his thought is not that the ideal deliberators should be replicated ('non-ideal agents are both imperfect and costly'). 'So the challenge is to find institutions that generate roughly the right results' (Kitcher 2001: 123). On particular issues, Kitcher, in fact, presents evaluations that he says, or suggests, would follow from an ideal deliberation. He claims, for instance, that ideal deliberators would probably 'view human cloning as a bizarrely complex way of achieving goals of relatively minor significance in comparison with the urgent health needs of the world's poor' (Kitcher 2011: 234). He also suggests that the ideal deliberators would consider it wise for a scientist to postpone the publication of a particular study on climate change to avoid giving ammunition to climate sceptics (Kitcher 2011: 184. See further examples in Kitcher 2011: 235, 239, 245. Compare Brown 2013).

One might doubt, of course, whether the ideal deliberators would come to the same conclusions as Kitcher does on these specific issues. It is clear, however, that Kitcher here offers, or at least suggests, a further approach to address the value-issues that a scientific advisor faces. The downsides of vulgar, democratic decision-making can be avoided by the introduction of a more enlightened democratic arrangement. However, that approach can also become a form of 'enlightened decision-making' by the scientist. (In the following two chapters, this option will be further investigated).

From One Elite to Another

In Kitcher's well-ordered science, as in Rawls' original argument, some real-world elements have been bracketed. One such element is power. The ideal settings are arranged so that rationality, other-regarding, and substantial dialogue predominate. In real-life situations, however, aspects of power and rent-seeking cannot simply be eliminated from the equation. That point is also true for deliberative arrangements. As studies in interactive or collaborative policymaking have shown, some participants manage to be more equal than others. They are more aware of the venues that exist, and they have the knowledge and the skills to dominate the agenda and the discussions. The Dutch Academy of Sciences did receive upon its already mentioned invitation to the public many well-elaborated research proposals of organized special interest (such as building and transportation companies) and of university research groups. In this case, it was quite clear that these agents were addressing their interests, but in other types of arrangements, that fact might not be so obvious. Vulgar democracy, according to Kitcher, is often dominated by small groups with intense interests (Kitcher 2011: 76). However, what guarantees that deliberative arrangements in such settings are not threatened by the same evil, an evil that might be even more effective and less visible?

Advocates of deliberative democracy occasionally suggest that individuals in an open debate can overcome points of dispute and develop common understandings or at least reach a consensus (Elster 1999). Kitcher is not that optimistic and acknowledges that disagreement on aims and values can be persistent; in such cases, majority vote should decide. This approach, however, might easily lead to domination by a particular majority choosing research topics and a consideration of risks that particularly fit its concerns and interests. The voices and needs of minority groups—groups that might also have problems in finding venues and lack skills—might be systematically neglected in research decisions (compare Dupré 2013). This point might also be described differently; a research agenda dominated by a certain societal majority will most probably contain few studies that are motivated by an ambition to unmask dominating (discursive) power, show hardships and neglect of 'invisibles', or emancipate minority groups. In summary, democratization of science might mean exchanging one type of elitism for another.

Innovative Capacity

A last type of complication of democratizing science that must be mentioned here is its relation to innovative research. Two dangers have been noted. Initially, as in other sectors, science democratization entails some level of bureaucratization. Systematically involving lay persons—in any form—means organizing deliberation, registration of decisions and motivations, and control of procedures and implementation (Tholen 2015). The large body of existing studies of bureaucratization suggests that this development is likely to have consequences that lead to a practice that differs from the ideal of a well-ordered science. A scientific practice steered by long winding democratic procedures might not be the most effective means to reach new solutions and techniques to address societal problems (Sarewitz 2011).

Furthermore, shifting the focus in scientific agenda-setting completely to democratic fora might easily lead to a concern for quick fixes of current societal problems and short-term research agendas. Many scientific findings only proved useful some time after they were discovered and in ways that had not previously been conceived—for example, new chemical substances proved effective medicine, or electrical phenomena that could be used in communication. Following leads in existing research programmes or scientific intuitions might bring about innovations that are blocked when research is merely driven by contemporary political concerns (Sarewitz 2011). On this issue, Kitcher's sophisticated democracy might not do much better than the vulgar democracy that he abhors.

4 Conclusion

Addressing the non-epistemic questions in research and advice democratically is an attractive solution. Given the societal consequences of these decisions—and, one might add, given that science, at least partly, is publicly funded—it appears reasonable to let the people decide somehow. Democratizing science, however, is not without complications and downsides. Democratization cannot be the sole solution to the issue of value-decisions in scientific advice, primarily because of the abundance of choices that must be made. The selection of the issues that will be made democratically and the approach to tutoring even make for new choices that a scientist, here, must make.

As Kitcher notes, democratization of science demands some type of representation, which might easily lead to the exchange of one elite for another. It can be argued that the common good might also be represented by the community of scientists—which might not be the worst scenario (compare Brown 2013). It would imply, at least, less need for the bureaucratic institutionalization that often accompanies democratization. It would also leave more room for innovative long-term research programmes.

There is a further reason why democratization does not and cannot take the burden of judgment from the shoulders of scientist advisors completely. Earlier (see Chap. 6) reference was made to the individual responsibility that scientists have because of their particular knowledge. In any particular democratic procedure, certain information might be neglected, and the severe consequences and risks of certain decisions might be misjudged. Such a result might also occur in democratic arrangements surrounding science. The responsibility of scientists might entail that they surpass their democratically governed institute's decisions and use channels to voice their concerns to the larger public. In any particular organization and setting, there might be (unintended) barriers to open deliberation, thereby causing that deliberation to deviate from an ideal democracy. The safe road of seeking publicity, a road that every citizen can adopt for a 'democratic appeal', can remedy such a shortcoming (compare Bovens 1998). There appears no reason to prevent scientists who believe they have a responsibility to inform the wider public from taking this civic road.

This variety of considerations implies that scientist advisors inevitably must make value-decisions. Bringing in democracy cannot (completely) solve all of the issues that arise in the course of research and advice. Kitcher, as mentioned above, offered a further reason for democratization. He observes that scientists often have a scientist perspective. They misunderstand the true nature of many value-issues and take them for epistemological ones. Kitcher is most probably correct in this observation (see on this point the introduction to Chap. 1). Democratization might be an effective means of making scientists aware of this misunderstanding. Democratization is, however, not the only approach to address this problem. More concern for this issue in science education might have the same, or even more, effect. We will return to this point later.

Realizing that until now, we have been considering what democracy can do for science and the scientist, a further question arises. The issue might also be reversed; what can and should science do for a democratic society? Above, democracy has essentially been understood as an (ideal) procedure for decision-making. However, what does it mean if we bring democracy into play not as a mere procedure but in a more substantial way? What does it mean for the decisions that scientist advisors must make if we acknowledge that they make these decisions in the context of a democratic society? Can there be guidelines for the decisions they must make, given their role in such a society? This question is one more issue we must return to below.

References

Bovens, M. (1998). *The Quest for Responsibility. Accountability and Citizenship in Complex Organisations.* Cambridge: Cambridge University Press.

Brown, M. B. (2004). The Philosophy of Science Policy. Review of Ph. Kitcher (2001) Science. *Truth and Democracy. Minerva, 42,* 77–95.

Brown, M. B. (2013). Review of: Ph. Kitcher (2011) Science in a Democratic Society. *Minerva, 51,* 389–397.

Douglas, H. (2004). Border Skirmishes between Science and Policy. In: P. Machamer & G. Wolters (Eds.), *Science, Values, and Objectivity.* Pittsburgh, PA: University of Pittsburgh Press.

Dupré, J. (2013). Review of Ph. Kitcher (2011) Science in a Democratic Society. *The Philosophical Quarterly, 63,* 408–410.

Elster, J. (1999). The Market and the Forum: Three Varieties of Political Theory. Essays on Reason and Politics. In J. Bohman & W. Regh (Eds.), *Deliberative Democracy* (pp. 205–242). Cambridge, MA: MIT Press.

Fischer, F. (1993). Policy Discourse and the Politics of Washington Think Tanks. In: F. Fischer and J. Forester (Eds.), *The Argumentative Turn in Policy Analysis and Planning* (pp. 21–42). Durham, NC: Duke University Press.

Flyvbjerg, B. (2001). *Making Social Science Matter: Why Social Inquiry Fails and How it Can Succeed Again.* Cambridge: Cambridge University.

Jennings, B. (1987). Interpretation and the Practice of Policy Analysis. In: F. Fischer and J. Forester (Eds.), *Confronting Values in Policy Analysis: The Politics of Criteria.* Newbury Park: Sage.

Kitcher, Ph. (2001). *Science, Truth, and Democracy.* Oxford: Oxford University Press.

Kitcher, Ph. (2011). *Science in a Democratic Society.* Amherst, NY: Prometheus Books.

McSwite, O. C. (1997). *Legitimacy in Public Administration: A Discourse Analysis.* Thousand Oaksm, CA: Sage.

Rawls, J. (1972). *A Theory of Justice.* Oxford: Oxford University Press.

Sarewitz, D. (2011). Looking for Quality in All the Wrong Places, or: The Technological Origins of Quality in Science Policy Advice. In J. Letsch &

P. Weingart (Eds.), *The Politics of Scientific Advice. Institutional Design for Quality Assurance* (pp. 54–70). Cambridge: Cambridge University Press.

Tholen, B. (2015). Citizen Participation and Bureaucratization: The Participatory Turn Seen through a Weberian Lens. *International Review of Administrative Sciences, 81*(3), 585–603.

Wagenaar, H. (2011). *Meaning in Action: Interpretation and Dialogue in Policy Analysis*. New York: M.E. Sharpe.

CHAPTER 9

Solving the Issue by Introducing Principles and Procedures

Abstract A third means of addressing the issues that were noted earlier uses principles and procedures. It uses (and often combines) elements from the utilitarian and deontological traditions to provide guiding principles for scientific advisors. This approach has its values but also must confront two sorts of problems. One sort of problem concerns the application of the principles; the other the choice between different principles.

Keywords Principles · Procedures · Utilitarian · Deontological

1 Introduction

In the previous chapters we found that scientific advisors must make value-decisions in all phases of their work; even democratization cannot lift this task from their shoulders completely. In this chapter and in the next, we will focus on approaches that acknowledge this role of scientific advisors. Both approaches emphasize that scientists must make many choices that are essentially social or ethical but not epistemological in character. They differ, however, in their answer to the question of how scientists should address them. In this chapter, an approach that focusses on principles and procedures will be considered.

B. Tholen, *Virtue and Responsibility in Policy Research and Advice*,
DOI 10.1007/978-3-319-65253-5_9

2 Rules for Scientific Advisors

Occasionally, an explicit endorsement has been presented to develop and use a principle-approach for value-decisions in science. A nice example is Callahan's call for 'well-established first principles [...] from which one can deduce the right decision in any specific instance' (Callahan 1973). His advocacy focussed on the decisions that must be made in biological research, but his message does not contain any particular reasons to limit the approach to this field.

Callahan himself did not provide particular principles. Others, however, have done so as a solution to value-questions. An example already encountered is Kitcher's proposal for accomplishing well-ordered science through the procedure of an imaginary meeting. His approach is clearly inspired by John Rawls' procedural method. Rawls (1972) describes an approach to normative reasoning that leads, so he maintains, to a generally acceptable and applicable notion of social justice. One must imagine a group of individuals, properly informed on relevant social, economic, and psychological regularities, deliberating and deciding together on basic principles. To guarantee that in this deliberation, real-life inequalities are not simply replicated, Rawls asks us to imagine that the participants have no knowledge of their particular position in society or of their own talents, worldview or preferences. In this state of partial blindness, participants must deliberate upon which principle would be most to their advantage. They will choose, Rawls argues, a particular set of rules of justice. Kitcher, like Rawls, uses a procedure that allows scientific advisors to find proper answers to the decisions they must make. He also lets us imagine a well-informed (tutored) group of persons that represent all people in society and that must decide, in several steps, on questions of science and advice (Kitcher 2001). As mentioned above, on several issues Kitcher suggests what he thinks the outcome of such sessions would be (for instance, on when to make findings public). Contrary to Rawls, Kitcher does not offer a specific elaboration of the arguments that might be used in the deliberations; nor does he present a particular set of principles that follow from the procedure. His argument, however, is a nice example of a procedural approach to addressing value-issues in science.

One of the principles that Rawls derives from his procedural approach is the *difference principle*. It expresses, essentially, a special concern for those persons who are worse off. Some scholars who have written on scientific value-choices have presented principles that express a similar concern. Popper, for instance, shows a parallel ethical consideration in his argument for doing science (Popper 1993). This concern for the weak and the deprived resurfaces in his comments about holistic policy-making based on dogmatic worldviews instead of piecemeal engineering based on a falsificationist orientation (Popper 2002). Jasanoff expresses a like understanding under the heading of 'technologies of humility'. This demands a focus on what can be done to make policies more inclusive and to help those who are harmed and marginalized (Jasanoff 2015: 1746). Neither Popper nor Jasanoff build on Rawls' argument; nor do they provide any other explicit ethical or philosophical rationale. However, they present on this point nice examples of scholars who offer particular principles for addressing scientific value-choices.

Further examples of such principles are the precautionary principle and the proactionary principle. Both are often invoked not only in policymaking but also in scientific advice. The *precautionary* principle comes in many formulations. Essentially, it implies a duty or mandatory action to refrain from introducing a new policy, instrument, substance, or approach if the degree of harm (if any) it might entail is uncertain. In simple terms, better safe than sorry (comp. Cranor 2004; Sandin 1999). The principle is often invoked not only in the field of environmental issues but also in medicine and health, data protection, genetic modification, and human enhancement. Particularly in the latter field in recent years, some have opted for a *proactionary* principle. Advocates of transhumanism (that is, developing and making available new technologies that enhance human intellectual, psychological and physical capacities) have endorsed it (e.g., Fuller and Lipinska 2014). The principle says that if there is no proof of immediate harm, it is appropriate to perform innovative experiments aimed at overcoming handicaps, hunger, diseases, and other hardships from which people now suffer.

Elliot suggests employing 'informed consent' as the guiding principle in scientific advice (Elliot 2011). He also proposes to use this rule beyond the areas in which it already has an established place, such as medical treatment and research. Following Scanlon, he explains that scientists who are guided by this maxim provide the knowledge that people need to pursue their life projects (Scanlon 1998 in Elliot 2011: 224). The scientific advisor contributes to the autonomous decision-making of others by ensuring that they have all of the relevant, scientifically established knowledge.

A different type of principle can be identified in arguments in favour of a critical approach in social sciences. The advocates often differ in the inspirational philosophies to which they refer, for example, *Frankfurter Schule*, Feminism, and Post-modernism. They also differ in the typical field of concern: for example, women and gender studies, colonialism and cultural studies, and governmentality and securitization. However, in all of these critical approaches, a similar principle for addressing value-choices in research and advice can be discerned; science should bring to light relationships of power and thereby reduce entrapment in (discursive) systems of domination and dependence (see, for instance, Bohman 2005; Box 2004; Farmer 2005). Science should provide people with the tools to reflect upon their position. It should be interested not only in finding knowledge to realize prevalent goals effectively but also in helping people emancipate themselves from social situations that appear normal or neutral but that in fact mean suppression of some to the advantage of others (Habermas 1971). The basic understanding, critical theorists maintain, is actually to follow up on Kant's *adagio* of Enlightenment and bring the knowledge that enables 'man to emerge from his self-incurred immaturity' (Kant 1991; Foucault 2007).

These examples of principles and procedures differ in many respects—in content and in the particular decisions in the process of research and advice they are intended to guide. However, they have some characteristics in common. First, in all of these examples, a procedure or rule is formulated that can be applied, whatever the particular issue or circumstance. Their formulation is always generic, not conditional. Herein they resemble algorithms in mathematics—a rule or procedure that, when carefully followed, brings the right result. Second, they are not provided as a mere suggestion or possibility among others, but as an ideal. They are presented as overshadowing any alternative and, therefore, evidently the best thing to do. Third, this presentation provides them the character of a guideline that is mandatory.

Together, these characteristics place the principles well within the modern ethical traditions of rule ethics (deontology) and utilitarianism. These ethical traditions resemble one another in their ambition to provide some context-independent procedure to decide which action is the better one. Rule ethics, as most prominently elaborated by Immanuel Kant, focusses on human dignity, on the respect for all persons, and on their ability to make their own decisions rationally (Kant 1998). Utilitarianism establishes formal guidelines and principles based on the intuition that sorrow, pain, and harm should be minimized as far as possible and that barriers to individual development should be abandoned (Mill 1991a, b).

The specific procedural and principled answers to the value-issues in science that were mentioned above, in fact, combine elements of both. The Rawlsian procedure that inspired Kitcher is designed so that individual worth and autonomy are considered (Kant), but all participants strive under partial blindness to the greatest advantage for each (Mill). In the precautionary principle, to provide another example, the ambition to reduce harm (Mill) is formulated as a mandatory principle (Kant).

3 Problems and Complications

Deontological and utilitarian theories provide elegant answers to ethical issues—general, simple principles and procedures. These answers, however, do have their particular complications both in the choice of a principle and in its application. To show these complications, I will examine more closely the examples that were introduced in the section above.

The Application of Principles

Kitcher presents a Rawlsian procedure to find answers to value-issues in science—an imaginary meeting of well-informed individuals. However, does Kitcher's description of this setting provide us with sufficient characteristics and constraints to arrive at any particular conclusion? Kitcher himself, as I mentioned in the previous chapter, appears to deduce certain answers from this procedure (for instance, on when to publish research findings on a politically controversial subject, such as climate change). In these instances, it is unclear, however, why the participants of the imaginary meeting would come to those particular conclusions (comp. Brown 2013). To be sure, in Rawls' original argument, the imaginary meeting is much more elaborated. However, even there it is doubtful whether the particular principles that Rawls himself endorses would be chosen (see for instance Daniels 1989). The proposed procedure thus appears too vague to lead to particular answers on particular issues.

The precautionary principle has been criticized for similar reasons. At first sight, the rule appears to be clear: 'if it is uncertain whether a new measure or approach will entail harm, do not introduce it.' However, how then should uncertainty or risk be understood? Should the precautionary principle be viewed as a barrier to all change when consequences are uncertain to some extent? This approach would mean a complete standstill, because we can never be completely certain about the consequences of our actions (on systematic limitations on prediction, see Popper 2002) for instance . The precautionary principle would be, as Sunstein (2002) states, a paralyzing principle. However, if we accept that some uncertainty is inevitable, where should the threshold be, what level (or type) of risk is acceptable and what is not?

A similar issue arises when we consider 'harm'. Does 'harm' mean material damage and costs to individuals and the public, or should immaterial goods also be considered? Is damage to the environment a matter of measurable pollution, or is it about diminishing biodiversity or no-longer-functioning biotopes? A more detailed judgment is clearly needed for the principle to be applied (Holbrook and Briggle 2014).

The basic concern of those who endorse the principle is avoiding or diminishing harm. However, is precaution in introducing new instruments and approaches always the best means to realize that result? Strict limits on the introduction of experimental, inadequately tested pharmaceuticals might protect people, but they also might prevent people from receiving the benefits of new medications. For example, blocking the introduction of new technologies means that existing ones can go on polluting, that hunger remains because drought-resistant genetically modified crops are banned, and that diseases flourish because food additives are not allowed. The precautionary principle thus does not always appear to lead to the result that it is said to serve (Sunstein 2005). This point would imply that the principle should not be applied categorically, and that a form of judgment that considers the particular context and consequences is more appropriate. If this proposition is true, however, then the precautionary principle must be viewed as a mere guideline and not as a mandatory rule (Mandel and Gathii 2006: 1072).

Elliot's principle of informed consent also contains a problem of application. This principle maintains that the scientific advisor should provide others with the knowledge that helps them to pursue their life projects best. The complication here lies in determining what is to be taken as someone's life plan. Should one pursue some expressed preference (such as 'I want to know where I can buy cigarettes'), or should one try to ascertain what the recipients of advice actually value (for example, their health or leading a pleasant life)? Translated to the field of policymaking, should scientists follow the explicit questions of the policymaker in their research and advice, or should they somehow try to determine what the policymaker really values? The latter will inevitably demand judgment and interpretation. To emphasize this point, we can refer to the position of critical studies. According to scholars in the critical tradition, people's understanding of their own position tends to be distorted. They might have particular life projects, but these projects are, in fact, results for dominating power constellations. For instance, women might have socially induced life plans that in fact limit their potential. Actually contributing to individual autonomous decision-making from a critical perspective means offering knowledge to overcome the current 'life' project.

The instruction of critical studies that research and advice should be emancipatory, however, can also be confronted with a comment concerning its application. The critical expression of a particular knowledge-interest cannot itself guide scientists in the specific decisions they must make. It does not tell them what dominating power must be exposed. Nor can it answer the question, what exactly does it mean to contribute as a scientist to a more critical and reflective attitude in particular groups. Does Weber's advocacy for presenting inconvenient truths count as a strategy along these lines? As we noted earlier, such an approach might lead to a value-reorientation in policymakers and others. The point here, again, is not to discredit the concerns that are expressed in this principle but to note that the principle itself does not appear to be sufficient to answer the issues that scientists encounter.

The principled approach promises an elegant simple algorithm to address value-problems in scientific advice. A first correction to this promise must be that the application of a principle or procedure always involves something more—further knowledge of particular circumstances and a capacity to interpret the principle for those particular settings. Thus, the rule might prove to be more a guideline than a strict law-like principle.

The Choice of Principles and Guidelines

A second type of comment centres on the observation that all principles and procedures express, in fact, a particular value and no other values. This comment questions the practice of always giving one value priority. We will examine the set of examples again.

In Rawls' imaginary meeting, all participants are understood to be mutually disinterested and choosing an outcome that is most beneficial for each as an individual (Rawls 1972). Critics have noted that Rawls' design is thus strongly biased towards individualism. The participants of the meeting are modelled as solitary persons, focussed on realizing their own preferences and life plans. A basic concern for the well-being of (particular) others is not programmed into this virtual meeting that, in consequence, gives individual liberties a prominent place in Rawls' political theory. Our concerns for collective values, communities, and civic associations are neglected, critics maintain (Sandel 1982). A similar point could be made concerning Kitcher's procedural approach. Central to his well-ordered science is the participation of citizens in the decision-making on values for science and advice. These citizens are tutored on the particularities of (the current standing of) science. The central focus is, however, on what citizens want from science, implying a bias of the procedure towards finding the issues and questions that society would like to prioritize on the scientific agenda. Long-term projects with no clearly specified (short-term) societal use would most likely receive only a very low place on the list of the imaginary deliberators. Toying with ideas, experimenting with not yet understood substances and phenomena, probing into ancient or remote subjects—that is, research activities not motivated by short-term usefulness and application—most likely would be activities with little appeal in the imaginary citizen assembly.

The precautionary principle has been challenged by advocates of alternative principles since its popularity began rising in the field of policy-making. One such alternative is the proactionary principle that was already mentioned above. Advocates of this latter principle mark the precautionary principle as conservative, leaving the burden of proof (of not being harmful) with those who want to introduce new approaches. They cheer the innovative character of the proactionary principle. Some observers of this discussion picture a disagreement between those who essentially distrust technology and those who believe in salvation by technology (Holbrook and Briggle 2014). Thus, it becomes clear that the precautionary principle expresses a particular value but neglects another. This point can even be taken one step further. Both the precautionary principle and the proactionary principle focus on (avoiding or repairing) harm in any form. One might be against the introduction of a new approach—for example, genetic modification in plants, animals, or even human beings—and the other might favour it. Both point to consequences—the good it promises against the harm it might entail. In focussing on the consequences, however, both miss a completely different type of value. In the case of genetic modification, that value might be expressed in the comment that we should not be playing God or that nature (as in the body of human beings) has an integrity of itself that should not be disturbed. Again, we encounter a principled perspective that expresses a different value from the precautionary principle—or the proactionary principle, for that matter.

This is not the place to delve deeply into the arguments underlying advocacies for a critical social science. (For a critique, for example, of Habermas' argument for a priority focus on emancipatory knowledge interests over other interests, see the contributions in the last part of Benhabib and Dallmayr 1995). Noting what it means to adopt the disclosure of dominating power structures as the central task for research and advice can, however, also illustrate the point here. A science that decides to study and realize emancipation must neglect other values. For instance, it must ignore how personal bases of power (different from structural or discursive power) are used, why they are used, and what consequences they have for the liberty of individual citizens and the functioning of governments and other organizations. It will also refrain from studies of, for example, the causes of famine, disease, and inefficient use of public means or on the effectiveness of all types of measures to address such problems. The point is not that there is no value in critical studies—on the contrary. However, other values also can guide research and advice, and it is not evident that these are always trumped by the value of emancipation.

The second type of comment upon the principled approach maintains, in summary, that principles and procedures express a particular value, whereas there are also other values, occasionally included in alternative principles. The issue then arises concerning which of the values (or principles) should receive priority. The examples given here strongly suggest that different settings might demand different valuations. The principled approaches themselves do not answer the question of how different values should be balanced or prioritized in particular cases. A procedural approach to evaluate different perspectives (for instance, the one that Rawls offers) in fact itself contains a particular prioritized value.

4 Conclusion

This chapter focussed on approaches that present principles and procedures to guide the value-decisions that scientific advisors must make. The examples given above of such approaches show a broad diversity of principles. What they all have in common, however, is that they try to offer simple and universally applicable guidance, just as a system of law does.

These approaches fit well within the modern traditions of deontological and utilitarian ethics. That quality, however, also makes them vulnerable to the critiques that these ethical approaches often must confront. First, the principles and procedures themselves lack precision and cannot lead to conclusive answers for particular issues. More is needed—in terms of particular knowledge and judgment—to be able to decide what approaches are required for particular cases. Second, the rules and procedures prioritize particular values. Other values are neglected or set aside, which does not always lead to a convincing result. This point is most clear in cases in which competing principles are endorsed. Critics have noted that procedural approaches that promise to provide an impartial answer to such a plurality of principles in fact are themselves already value-biased.

Principled and procedural approaches present important insights and strong elaborations of our moral intuitions; eventually, however, they cannot live up to their promise. Commentators have argued that principles and procedures might better be taken as guidelines and not be treated as decision procedures. They can be of help in our deliberation, but cannot replace it (e.g., Mandel and Gathii 2006; Holbrook and Briggle 2014). For an alternative approach to value-decision-making in the field of scientific advice, some refer to a pre-modern ethical tradition—virtue ethics (e.g., Macfarlane 2009).

References

Benhabib, S., & Dallmayr, F. (Eds.). (1995). *The communicative ethics controversy*. Cambridge, MA: MIT Press.

Bohman, J. (2005). Critical theory. *Stanford encyclopedia of philosophy*. Retrieved March 15, 2017, from https://plato.stanford.edu/entries/critical-theory/.

Box, R. C. (2004). *Critical theory in public administration*. London: Routledge.

Brown, M. B. (2013). Review of: Ph. Kitcher (2011) Science in a democratic society. *Minerva, 51*, 389–397.

Callahan, D. (1973). Bioethics as a discipline. *Hasting Center Studies, 1*(1), 66–73.

Cranor, C. F. (2004). Toward understanding aspects of the precautionary principle. *Journal of Medicine and Philosophy, 29*(3), 259–279.

Daniels, N. (Ed.). (1989). *Reading Rawls: Critical studies on John Rawls' A theory of justice.* New York: Basic Books.

Elliot, K. C. (2011). *Is a little pollution good for you? Incorporating societal values in environmental research.* Oxford: Oxford University Press.

Farmer, D. J. (2005). *To kill the king. Post-traditional governance and bureaucracy.* Armonk, NY: Sharpe.

Foucault, M. (2007). What is enlightenment? In idem: *The politics of truth* (pp. 97–120). Los Angeles: Semiotext.

Fuller, S., & Lipinska, V. (2014). *The proactionary imperative: A foundation for transhumanism.* London: Palgrave.

Habermas, J. (1971). *Knowledge and human interest.* Boston: Beacon Press.

Holbrook, J. B., & Briggle, A. (2014). Knowledge kills action: Why principles should play a limited role in policy-making. *Journal of Responsible Innovation, 1*(1), 51–66.

Jasanoff, S. (2015). Serviceable truths: Science for action in law and policy. *Texas Law Review, 93,* 1723–1749.

Kant, I. (1991). An answer to the question: What is enlightenment? In H. S. Reiss (Ed.), *Kant's political writings* (pp. 54–60). Cambridge: Cambridge University Press.

Kant, I. (1998). *Groundwork of the metaphysics of morals.* Cambridge: Cambridge University Press.

Kitcher, Ph. (2001). *Science, truth and democracy.* Oxford: Oxford University Press.

Macfarlane, B. (2009). *Researching with integrity: The ethics of academic inquiry.* NewYork: Routledge.

Mandel, G. N., & Gathii, J. T. (2006). Cost-benefit analysis versus the precautionary principle: Beyond Cass Sunstein's laws of fear. *University of Illinois Law Review, 5,* 1037–1080.

Mill, J. S. (1991a). On liberty. In idem; *On liberty and other essays* (pp. 5–128). Oxford: Oxford University Press.

Mill, J. S. (1991b). Utilitarianism. In idem; *On liberty and other essays* (pp. 131–201). Oxford: Oxford University Press.

Popper, K. (1993). *The open society and its enemies* (Vol. 2). London: Routledge & Kegan Paul.

Popper, K. (2002). Science: Conjectures and refutations. In idem, *Conjectures and refutations* (pp. 43–78). London: Routledge.

Rawls, J. (1972). *A theory of justice.* Oxford: Oxford University Press.

Sandel, M. (1982). *Liberalism and the limits of justice*. Cambridge: Cambridge University Press.

Sandin, P. (1999). Dimensions of the precautionary principle. *Human and Ecological Risk Assessment, 5*(5), 889–907.

Scanlon, T. (1998). *What we owe to each other*. Cambridge, MA: Harvard University Press.

Sunstein, C. R. (2002, Winter). The paralyzing principle. *Regulation* (pp. 32–37).

Sunstein, C. R. (2005). *Laws of fear: Beyond the precautionary principle*. New York: Cambridge University Press.

CHAPTER 10

Solving the Issue by Bringing Virtue Ethics into Play

Abstract A fourth approach to addressing the social value issues follows the virtue ethical tradition. Two approaches are presented: one that starts from typical scientific virtues and another that focusses on practical wisdom (phronesis) that is at home in politics. A third approach that builds on MacIntyre's notion of practices is introduced to integrate the two, leading to the idea of the academic citizen.

Keywords Virtue ethics · Virtue · Scientific practice · Academic citizen

1 Introduction

This chapter, like the previous one, explores the possibilities that a particular ethical tradition must offer. Here, the focus is not on procedures and principles that must guide actions but on virtues, that is, on attitudes and orientations of persons. The virtue ethical tradition has its roots in the philosophies of Plato and Aristotle. Over the last half-century, particularly since the seminal *After Virtue* by MacIntyre (1985), this ethical approach has developed into a central contemporary ethical theory, next to utilitarianism and deontology.

Moreover, virtue ethics has been employed in recent years in epistemology and science studies. Here, several such approaches will be

B. Tholen, *Virtue and Responsibility in Policy Research and Advice*,
DOI 10.1007/978-3-319-65253-5_10

discussed. First is an approach that focusses on science as a practice that flourishes when its participants enact particular virtues. The second section addresses a virtue ethical approach that focusses on one particular personal competence—*phronesis* (practical wisdom). Both approaches are promising but leave essential issues unanswered. In the last section, therefore, a further elaboration will be suggested, following the ideas of MacIntyre.

2 Virtues of Scientific Practice

In recent years, scholars have developed new approaches to acquiring true beliefs and reliable knowledge building on virtues (e.g., Greco 2010; Sosa 1980; Williams 2002). Trying to solve problems that haunt principled epistemologies (in particular, coherence and foundationalist approaches), they have turned to attitudes of scientists and asked what individual characteristics in researchers bring reliable knowledge. Although some contributions remain on a meta-ethical analytical level (Greco 2010), others have tried to note the relevant scientific virtues. Roberts and Wood (2007), for instance, explain how love of knowledge, firmness, courage, humility, authenticity, and generosity are of importance. Macfarlane (2009) presents a similar list, and in many examples, he shows how they function. He illustrates, for instance, how courage is necessary in daring to probe new types of research, asking new questions, and crossing disciplinary boundaries. He shows how resoluteness is necessary to keep on searching when results come slowly and tend to disappoint. Finally, to provide a last example, he notes how sincerity is at odds with making up research findings or the trimming of data.

Virtues, these advocates explain, must be understood as well-developed individual attitudes. These are not merely inborn talents; nor are they the effect of a lucky hand at any particular moment but rather the result of training or schooling, just like the skills of a carpenter or a well-coached sportsman. A virtue is a relatively stable competency and involves knowing how to act properly in particular cases. Virtues, furthermore, can be understood as means between two extremes. Courage, for instance, is the virtue between cowardice and recklessness (Pellegrino 1992; Roberts and Woods 2007; Macfarlane 2009). Advocates typically refer to Aristotle as a founding father of virtue ethics. Aristotle defined virtue as 'a settled disposition of the mind determining the choice of the actions and emotions… as the prudent man would determine it'

(Aristotle 1982: 1106b36–38). This disposition or attitude can be developed by moderating extreme emotions. One can become a more courageous person, for instance, by tempering one's emotions of fear and rashness.

Aristotle's definition of virtuousness as 'as the prudent man would determine' appears to be tautological. However, according to this approach, what is virtuous in any specific setting cannot be determined using some rule or algorithm in advance or beyond that setting. The specifics of each particular case and all of the relevant values and consequences must be considered. A person can develop his 'virtuosity' in a particular setting. MacIntyre explains this idea by introducing the concept (or re-introducing it, given its roots in Aristotle's ideas) of a *practice*. A practice is 'any coherent and complex form of socially established cooperative human activity' (MacIntyre 1985: 187). Examples of such a cooperative venture are a sports game, architecture, farming, but also scientific research. By engaging in any of such activities, one might obtain prestige, status, and money. MacIntyre calls them *external goods* because they can be realized or gained through participation in any activity. In fact, *real* engagement in the activity is not even necessary. A little cheating in for example a sports game or a scientific study might do the trick. Realizing the good that is *internal* to a practice, however, demands real involvement. What real involvement is depends upon the activity. For example, it means (and demands from the participants) something quite different in health care from in architecture. External goods can be an individual's property and possession. They are typically the object of competition and are a zero sum game. Internal goods, however, constitute an achievement for the whole community. Examples are knowledge, health, or art. One's virtues are the particular individual qualities that enable one to realize or contribute to the internal good of a practice (MacIntyre 1985: 188).

It is by participating within a practice that someone can acquire appreciation for the internal good and develop the relevant virtues. MacIntyre provides the analogy of a child who by the promise of some confectionery is tempted to give the game of chess a try and who, over time, if properly tutored, might start to appreciate playing the game without the reward of any external good. Similarly, students can be won over to become actively engaged in the practice of scientific research. Participants can further develop their virtues in the practices by experimenting, by learning from mistakes, but also from following the examples of those who show more experience or virtuosity in this type of activity. Nor are the practices themselves fixed settings. Sports, healthcare, science—they all have a history and evolve.

The virtues that Macfarlane, Roberts and Woods, and others present are those that guide scientists in realizing true knowledge and students learn when they are introduced into this practice. The virtues mentioned, however, are particularly focussed on epistemological questions. They guide towards reliable and robust statements about what the empirical world is like. They are of little help in addressing the decisions that demand some type of social or political valuation such as those discussed in Chaps. 3, 4, 5, and above.

Macfarlane acknowledges that those virtues have limited relevance. In an article on 'academic service and citizenship', he presents a further set of virtues that do concern the relationship between science and society. He mentions as principal virtues in this domain collegiality, loyalty, guardianship, benevolence, and engagement. They all, particularly engagement, express a role and responsibility for the scientist in the larger community (Macfarlane 2007, 2009: 164). Linking the scientist's value-decisions to his citizenship is a promising suggestion for our investigation. Macfarlane, however, focusses in his article on citizenship and engagement of scientists in a quite literal sense—becoming members of public committees and the like. The questions identified in this volume are not addressed. How might such issues, that appear to transcend the boundaries of scientific practice, be understood from a virtue ethical perspective?

3 Phronetic Science

A second virtue ethical approach chooses a different starting point. It argues that much of contemporary (social) science and scientific advice has a predominant rationalist or instrumentalist character. This rationalistic orientation considers policymaking and delivering scientific input for policymaking a mere technical exercise. What it essentially misses, the critics argue, is an understanding of the value issues that are involved (compare Chaps. 3, 4, 5, and above) and of the need for a means of answering them that does not follow a general, logical, rational scheme (e.g., Sanderson 2002; Flyvbjerg 2001). These critics try to make scientists aware of 'the rationalist fallacy … of interpreting policymaking as a purely intellectual exercise' (Majone 1989: 146).

In the alternative approach that these commentators present, they build on Aristotle's distinction between types of knowledge, more particularly between *episteme* and *phronesis*. *Episteme* is related to scientific knowledge; it concerns knowledge of what is invariable and true independent of context. *Phronesis* might be translated as having good judgment, prudence, or practical wisdom. It involves practical knowledge and expertise on how to behave in and address particular circumstances. *Phronesis* is about knowing what works, that is, how certain aims can be best realized. In that sense, it resembles what Aristotle called a *techne* (the ability of a craftsman). However, it is also about judging what aims might be best for these circumstances, given the effects of measures and given the more encompassing ends we are striving for (Aristotle 1982). *Phronesis*, like the virtues we discussed earlier, is a type of knowledge that cannot be learned from books; experience is crucial to develop it.

The comment that (policy) science and scientific advice should be more aware of value-issues and, therefore, turn to practical wisdom has been made by a variety of scholars in Public Administration (e.g., Lynn 1996; Majone 1989; Schön and Rein 1994; Vickers 1995). In recent research, however, this type of advocacy has been explicitly linked to Aristotle and virtue ethics. Flyvbjerg, for example, opts for a *phronetic social science*. This type of science is led by three basic questions, he maintains: Where are we going? Is it desirable? What should be done? (See esp. Flyvbjerg 2001: Chap. 5). Flyvbjerg concludes his book by presenting an elaborated example of what it means to practise phronetic social science. The research case concerns a city planning process. Researcher themselves try to become acquainted with the particular policy dispute to the best of their abilities. They obtain their own desks in one of the relevant organizations, work to become familiar with the background and history of the problem, and try to learn about all of the relevant actors and their preferences and power bases. They are involved in the process, for instance, in the construction of deliberation and negotiation panels and, eventually, are invited to take seats on such a panel (Chap. 10).

Remarkable in this example is that the scientists appear to have become one more agent in the policy process. Alternatively, given that they have no particular interest in the planning process at hand, they appear as administrators or aldermen who try to organize a deliberative process among stakeholders leading to an acceptable conclusion. Phronetic scientists themselves, in fact, appear to have become phronetic policymakers. (Flyvbjerg recognizes that the line has become quite thin: p. 161). Such a form of role taking can also be discerned in other advocacies of scientific policy advice as practical wisdom. The policy scientists are urged to try to understand the value complexity in the fields they advise on to consider multiple instruments but also reflect upon ends (Lynn 1996; Majone 1989; Schön and Rein 1994; Vickers 1995). Scientists are in fact invited to imagine themselves being the policymakers and even take active roles in the policy process itself. The idea that policymaker and researcher are in fact identified is strengthened by the comments of advocates for this approach on science's ability to realize any (relevant) theoretical knowledge (Flyvbjerg 2001: Chap. 3; Lynn 1996). Here, ironically, the comment that science mistakes policymaking for doing rational science appears to have reversed into reducing policy science to value-aware policymaking.

Phronesis might be important for scientists and scientific advisors, but the virtue ethical approach in this second form appears to lose the distinction between the role of the scientist and that of the actual policymaker.

4 Scientific Practice and Political Practice

The virtue ethical approach might be developed further on this point if we consider how MacIntyre has elaborated both science and politics. MacIntyre emphasizes that there are many different practices in society: in sports, arts, healthcare, science, and many more fields in which people cooperate. Moreover, the political community, if it is to be more than an instrument for individuals to realize profits and status (that is, external goods), must be such a practice, he maintains. The political practice has its virtues: justice, courage, and honesty. These are the most general (or basic) virtues because they inevitably also have a role in all other practices. They are relevant for all human interaction (MacIntyre 1985: 191). As is true for all other practices, the internal goal of politics as a practice can only be indicated in general terms—realizing the common good (compare health in the medical practice or knowledge in science). Essential to the virtue ethical approach, however, is that in a political community as a practice, all members can develop their talents or potential. Politics as the practice that rules over all other activities in society, therefore, is about directing all of those other practices so that individual development is best guaranteed (MacIntyre 1999).

A scientific policy advisor must be understood, therefore, as a member of two practices: the scientific one and the political one. It is in the issues that were identified in Chaps. 3, 4, and 5 that the practice of science and the political practice touch each other. Scientists who must address these non-epistemic value-issues must practise, in Macfarlane's words, academic citizenship. Although scientists, they must employ the typical political or civic virtues of justice, engagement, and *phronesis*. As academic citizens, they can contribute to public debate in a way that others cannot, offering robust, epistemologically certified knowledge. They can enhance the debate in a particular way. At the same time, they must choose on which particular forum they makes their contribution, on what subject, guided by what concern and in what form. In having to make these choices, they do not differ from other citizens.

What, then, precisely is the relationship between the two practices? MacIntyre follows Aristotle in the understanding that politics rules over all other practices. That idea must not be misunderstood. Although politics has some regulating role in placing other activities in society 'in their place', these practices do have their autonomy. In fact, they can only fulfil their role for the common good if they can be self-guiding to some extent. To use one of Aristotle's examples, a saddle might be a useful means for a general; however, a general best not try to instruct saddle makers on their craft (Aristotle 1982). This example illustrates that the relationship between science and politics is a somewhat complicated one. Science must have its own autonomy to flourish as a practice and to continue being able to provide knowledge. It would not do so if outsiders were to decide how to do research properly. The typical epistemological virtues of the research practice that Roberts and Wood (2007), Macfarlane (2007) mentioned must be guiding. Furthermore, science has the task of providing knowledge for the political community. What this statement means, however, is more complicated than in the case of the saddle for the general. First, the recipient is not simply a person but the rather abstract political community or, to be more precise, the public debate in the political community on how to realize the common good. Second, knowledge is typically not simply an instrument to realize a predefined end. Knowledge might, in fact, change and redefine ends. This point is encountered repeatedly in Chaps. 3, 4, 5, and above and is emphasized by the advocates of *phronesis*.

5 Summary

In this chapter, the focus was on virtue ethical answers to the value issues in science and scientific advice. I have presented two approaches both of which are promising but which, upon closer inspection, proved to be wanting. In the last section, an elaboration was tried that, following the line of MacIntyre, understands the earlier two approaches as complementary. According to this third position, it is particularly in the moments that non-epistemological value-questions arise that the scientific practice and the political practice touch each other. This interaction leads to the idea of the academic citizen for whom purely scientific virtues are insufficient but who also needs the virtues of a good citizen properly to address the value-issues that arise in all three phases of doing research.

Virtue ethics acknowledges that scientific policy advisors themselves have to address the value-issues; they cannot avoid individual judgment and rely on fixed rules and procedures or refer issues to democratic institutions. If the practices within which they are active are well-functioning they can develop their judgment-skills and employ them. The virtue ethical approach provides, therefore, an articulation of the actions and the virtues that a good scientific policy advisor should aim for. It does not provide, in fact it argues that there cannot be given, general answers, procedures, or rules on how to address issues relating to social values.

References

Aristotle. (1982). *The Nicomachean ethics.* Cambridge, MA: Harvard University Press.

Flyvbjerg, B. (2001). *Making social science matter: Why social inquiry fails and how it can succeed again.* Cambridge: Cambridge University Press.

Greco, J. (2010). *Achieving knowledge.* Cambridge: Cambridge University Press.

Lynn, L. E. (1996). *Public management as art, science and profession.* Chatham, NJ: Chatham House Publishers.

Macfarlane. (2007). *The academic citizen: The virtue of service in University Life.* Abingdon: Routledge.

Macfarlane, B. (2009). *Researching with integrity: The ethics of academic inquiry.* New York: Routledge.

MacIntyre, A. (1985). *After virtue: A study in moral theory.* Notre Dame, IN: University of Notre Dame Press.

MacIntyre, A. (1999). *Dependent rational animals. Why human beings need the virtues.* Peru, ILL: Carus Publishing Company.

Majone, G. (1989). *Evidence, argument, and persuasion in the policy process.* New Haven: Yale University Press.

Pellegrino, E. D. (1992). Character and ethical conduct of research. *Accountability in Research, 2,* 1–11.

Roberts, R. C., & W. J. Wood. (2007). *Intellectual virtues: An essay in regulative epistemology.* Oxford: Clarendon Press.

Sanderson, I. (2002). Making sense of 'what works': Evidence based policy making as instrumental rationality? *Public Policy and Administration, 17*(3), 61–75.

Schön, D. A., & Rein, M. (1994). *Frame reflection. Towards the resolution of intractable policy controversies.* New York: Basic Books.

Sosa, E. (1980). The raft and the pyramid: Coherence versus foundations in the theory of knowledge. *Midwest Studies in Philosophy, 5,* 3–25.

Vickers, G. (1995). *The art of judgment. A study of policy making.* Thousand Oaks, CA: Sage.

Williams, B. (2002) *Truth and truthfulness.* Princeton: Princeton University Press.

CHAPTER 11

Conclusion: Scientific Policy Advice and Ethical Judgment

Abstract It is concluded that science and scientific policy advice inevitably involve making decisions that involve values—epistemological and non-epistemological values. A virtue ethics that includes the strong elements of the other three solutions might show the best way to address these value issues. This approach urges a particular academic curriculum and leadership. It demands virtuosity in employing an academic citizenship and a continuous balancing between the memberships of two practices.

Keywords Scientific practice · Political practice · Responsibility Academic citizen · Education · Leadership

1 Introduction

The central question of this book is ***what type of value-decisions must expert advisors make, particularly in the fields of Policy Analysis and Public Administration, and how can they address those issues***? The answer to the first part of this question was formulated in Chap. 6; it will be summarized in Sect. 2. In the last four chapters, a series of approaches for addressing these value-issues has been discussed. Section 3 elaborates to what extent these four approaches can help us to address these value-issues. In Sect. 4 the contributions of a virtue ethical approach are further specified. The last section is dedicated to the particular

B. Tholen, *Virtue and Responsibility in Policy Research and Advice*,
DOI 10.1007/978-3-319-65253-5_11

complications that scientific policy advisors can encounter when they try to follow the ideal of the *academic citizen*.

2 Value-Choices in Scientific Advice

The first part of this book made clear that it is a misunderstanding to think that science and scientific advice can and must be free of value and value-judgments. First, science can only function to the extent that scientists follow methodological and logical rules and guidelines and employ the epistemological virtues. For example, if interpretations and causal theories are not seriously tested, data are made up or trimmed, or scientists are not honest about the procedures they followed, then science cannot provide valid statements about how phenomena in the world can be understood and explained.

Of course, the relevance of these epistemic values is not actually contested. By 'a science free of values', most in fact mean a science that is guided only by epistemological values and not by social, ethical or political ones. Max Weber is often invoked as the champion of this position. We examined more closely the different phases of science and scientific advice and observed that this interpretation of value-free science must also be denounced. Developing research, performing the study, and giving advice are activities that all involve many decisions that cannot be made by relying on epistemological values alone. Scientists must determine whether they merely want to show what works or to assist in developing a framework to understand a problem better. They must make choices such as between being subservient or acting as a Socratic gadfly, accepting a greater or smaller risk of making an error of the first category, focussing on innovative measures or being more conservative, adopting dominant value positions as leading or being critical, considering material consequences or also immaterial ones such as for example civic trust, and accepting the implicit focus on effectiveness in causal research or also including other values.

Values in science, as well as epistemological and social values, must not be considered deplorable external intrusions into a mere technical enterprise. Both types of values are, in fact, basic to science. Some might find this conclusion a reason for concern. Is society, given the potential impact of social value-choices in science on policymaking, not left at the preferences and whims of scientific advisors? This question brings us to the subject of the second part of this book; what would be a good way for scientists to address the non-epistemological value-decisions that they must make?

3 Addressing Social Value Issues in Scientific Policy Advice

An approach to addressing research advisory issues that is regularly suggested in Policy Sciences focusses on problem types. According to this approach, which was discussed in Chap. 7, problems can be characterized according to a certain grid. Each particular problem must be addressed in a particular way—for example, technical or critical, focussed on expanding or excluding options. An attractive aspect of this approach is that it considers the diversity of settings and extends beyond a simple one-size-fits-all solution. This scientific approach has its problems, however. It implies that one can objectively determine what in any given context the problem is. That idea is at odds with the finding that problem definitions (in the political field) are themselves objects of debate. Moreover, public understandings of problems might change as a consequence of scientific input. The problem approach exemplifies that a scientific approach cannot solve the value-issues. A second approach, discussed in Chap. 8, acknowledges that value-decisions must be made in science and scientific advice. Its answer is that such issues should be solved democratically. It maintains that, because scientific decisions have consequences for the public debate, and for large groups, or all citizens, some way of including everyone's interests and points of view is appropriate. In response to advocates of this solution, it has been noted that it is impractical and even impossible to leave all decisions to a citizen forum. Some decisions inevitably must be left to scientists, and scientists must decide which issues they present in what way to the public and how they tutor the public about them.

The conclusion that follows from both chapters is that we must turn to individual ethics for scientists to address the social value-issues. In Chap. 9, approaches were investigated that follow the modern ethical traditions of deontology and utilitarianism and seek solutions in procedures, rules, and principles. These approaches encompass advocacies for decision procedures, for principles such as the precautionary principle or informed consent and for critical emancipating research. Two critical comments arose in the discussion of this approach. First, the principles alone hardly provide sufficient guidance for addressing particular issues; further interpretation and judgment is necessary. Second, even principles we feel most strongly about, might in certain circumstances give way to some alternative principle. Individual judgment on what action is most appropriate for a particular situation is inevitable. Chapter 10 focussed on another ethical tradition, virtue ethics. In the field of science, scholars have in recent decades tried to revitalize this classical approach to ethical issues. Their contributions lead us to the understanding of scientific policy advisors as participants in two practices: a scientific practice that focusses on truth and that demands the realization of epistemic virtues and a political practice that centres on realizing the common good and has its own typical virtues. This virtue ethical approach helps to clarify the position of the scientific advisor as an academic citizen.

4 Virtue Ethical Elaborations

Virtue ethics can also help elaborate another point. It shows how guidelines can be of relevance, although they cannot provide the kind of mathematical certainty that principled approaches would like them to. From a virtue ethical perspective principles can be understood as guidelines or rules of thumb. Experienced participants in a particular practice can find that a particular consideration returns in many settings and that it often must be preferred over others. Such an observation might lead to the formulation of a principle such as the principle of informed consent. Within the virtue ethical practice, this principle is not, nor can it be, some algorithm that one can simply apply in every case. Its application demands a well-developed competency of judgment. That requirement involves knowing when the principle must be considered and what, in a particular circumstance, might count as 'being informed' and 'having consented'. Virtue ethics adds to the principled approach the realization that principles need interpretation. In other words, whereas virtue ethics emphasizes the need to be situation aware and to be trained in good judgment, the rule-approaches formulate basic principles and concerns that in many situations must be considered and that can be used to organize deliberation. One might conclude that virtue ethics and the principled approaches can be understood, to some extent, as complementary.

Virtue ethics can help, furthermore, to obtain a better understanding of what it means to be a responsible scientist. In Chap. 6, the responsibility of scientific advisors was elaborated, following the work of Thompson (1983), Douglas (2009). It was established that scientist advisors can and should be held accountable for their actions. When their actions, or a lack thereof, have led to harm and suffering, they are ethically accountable and deserve blame. The virtue ethical approach enables us to note another more encompassing notion of the responsibility of the scientific policy advisor. Responsibility, here, means more than being accountable for wrongs and harm. It means having a task, being entrusted with the care for the functioning of a particular practice. Scientists as members of the scientific practice have a responsibility for the functioning and flourishing of it. That responsibility consists of not only refraining from for example fraud and data trimming but also correcting others and fighting (financial and organizational) mechanisms that undermine realizing an internal good—finding truth. As members of the political practice, they have a responsibility for the realization of the common good, for instance in a concern for a well-informed public debate.

A last, and not least important contribution of virtue ethics is its understanding of and focus on ethical education and personal development. Virtue ethics emphasizes that the ability to judge in particular cases demands experience and training. Only by participating in practices and by practising and learning from others who are more experienced, can individuals develop the relevant skills. The institutions in which practices have their place must therefore be organized to contribute optimally to the education in virtuosity of the participants. For scientific practice, that means virtuosity not only in addressing epistemological issues (research techniques and methods) but also in addressing social and political issues. Such an education goes beyond merely presenting overviews of research traditions (such as positivism or the *Frankfurter Schule*). Dealing with non-epistemological issues in research and advice demands that researchers develop an understanding of the common good, that is the values and aims that are relevant in political practice, and an ability to make their own judgments on these issues. Accomplishing this goal involves a university curriculum in which these issues have their place. It also involves a type of academic leadership that stimulates open debate and taking responsibility (comp. Resnik and Elliot 2016; Tholen 2011).

5 A Member of Two Practices

The virtue ethical approach enables us to elaborate the understanding of scientific policy advisors as academic citizens—members of two practices. This understanding does justice to the multiple value set that characterizes the 'sophisticated value-free scientist'—epistemic values that guide some issues and social values for others. Following the first makes them good scientists as truth-seekers; following the second makes them good scientists and responsible citizens. Thus far, this mixed character of the academic citizen has been presented as unproblematic; its two sides have been regarded as complementary. Trying to live up to the ideal of the academic citizen can, however, be a complicated task. One source of complication lies in the possibility that both sides might make for different demands. Earlier chapters in this book provide some examples:

- Scientists might have to choose between performing a study that helps to develop further a research programme and a study that might (quickly) lead to an answer to a societal question or problem.
- Scientists might have the choice between measuring the value of their findings in terms of epistemological robustness and in terms of its practical usefulness.
- A scientist might have found results that are highly relevant for scientific discussion but that would, if published, most likely be used in political debate to fuel an opposition that they think undesirable (for instance, to question climate change).

What is at stake here is that the scientist must choose between being a good truth-seeker and being a good citizen. These two sides might not always be easy to reconcile. The scientific advisor not only is a member of two practices but also must match the demands of each. Occasionally, these demands conflict; in these cases, clever solutions that serve both sides are required. In specific cases, an uncompromising choice for one side must be made. In any case, conscientious deliberation appears to be demanded—we would probably not find it appropriate if the scientist simply flips a coin to come to a decision.

The duality in the position of the academic citizen can also be a difficult one in another respect. Scientists are at once like and unlike any other citizens. Their contribution to the political debate (ideally) has a more solid empirical base than other types of input, but their contribution is guided by social values, as is that of others. This ambiguity might easily be misunderstood. Citizens that witness the value aspects of scientific advice might question the special contribution that science can offer: 'it is just another political position expressing a particular interest'. Scientific advisors might try to escape from this complicated and threatened position. One approach is neglecting the social value side of decisions in science, understanding their trade as merely a technical one. 'I just do research, I strictly follow scientific method, I am neutral.' The previous chapters have shown that such a statement would misrepresent what one actually does. Answering critics by referring to one's integrity as a neutral researcher, as the scientists cited at the beginning of Chap. 1 did, is not an convincing response.

A second type of attempt to escape from this position is engaging in policymaking but without taking the role of an expert that has particular empirical insights to offer. Presenting oneself as a participant, similar to others involved in a policy process, is a strategy that is suggested in advocacies for participatory policy processes, as we saw earlier. The downside of this approach, of course, is that it no longer offers any input based on scientific research.

Living up to the demands of being a scientific advisor, instead of trying to escape from it, means being aware that one must aim at being objective (in the epistemological science) but accept that one cannot be neutral (in terms of social value choices) and must make judgments. One must try to be a good scientist *and* a responsible citizen.

6 Conclusion

Currently, moral education and value leadership are clearly a matter of concern in academia. The focus in these activities now, however, is on epistemological issues: for example, eradicating fraud, trimming data, and the appropriateness of data-mining techniques. Our conclusion is that a complete 'value-curriculum' for scientific education also involves the social value issues. Scientific policy advisors inevitably must make all types of non-epistemic value-decisions in all phases of their work.

Addressing this reality properly means that the scientist and policy advisor also must develop good judgment on these issues. In academic curricula, students must therefore learn to recognize social value-issues in research and advice. Courses in social and political philosophy and ethics are as important as courses in research method and epistemology. Being able to make judgments on social value-issues in research, virtue ethics emphasizes, demands actual discussion and training, on ethical, social, and political issues within the practice of science and advice.

References

Douglas, H. E. (2009). *Science, policy, and the value-free ideal.* Pittsburgh, PA: University of Pittsburgh Press.

Resnik, D. B., & Elliot, K. C. (2016). The ethical challenges of socially responsible science. *Accountability in Research, 23*(1), 31–46.

Tholen, B. (2011). Public virtue approaches. In P. S. Kim & M. S. de Vries (Eds.), *Value and virtue in public administration* (pp. 33–46). Basingstoke: Palgrave Macmillan.

Thompson, D. (1983). Ascribing responsibility to advisors in government. *Ethics, 93*(April), 546–560.

Epilogue

As mentioned earlier, this book is not intended as a textbook or critical study on Weber. However, given his prominent role as an example in this book's argument, it appears appropriate to end with a last reference to Weber's work.

In many of his writings, Weber addresses the position of the individual who conscientiously, in demanding situations, tries to find the right path. In his writings on religion, for instance, he explicates the difficulties of combining the ideals of religion and more worldly demands (e.g., Weber 1948). In *Politics as a Vocation*, he pictures the political leader as a tragic hero having to choose between not accomplishing results or doing so by employing evil measures. The position of the scientific advisor occasionally might also be a difficult one, particularly if the two hearts in the breast of the academic citizen do not concur, for example, in cases in which scientific and social responsibility make different demands.

There can be many reasons to neglect the value-issues in scientific policy advice. They distract from the methodical puzzles, they are sometimes very difficult to deal with, and they concern decisions that others might question. Suggestions to sidestep individual judgments—follow conventional rules and practices, leave the decision to a democratic institution—might for these reasons be welcomed. In such difficult cases the true academic citizenship of scientific policy advisors demands the type of courage that Weber envisioned in all real vocations: the courage to deal virtuously with all the responsibilities that come with the trade.

B. Tholen, *Virtue and Responsibility in Policy Research and Advice*,
DOI 10.1007/978-3-319-65253-5

Refernces

Weber, M. (1948a). Religious rejections of the world and their directions. In H. H. Gerth & C. W. Mills (Eds.), *From Max Weber. Essays in sociology* (pp. 323–359). London: Routledge & Kegan Paul.

Weber, M. (1948b). Science as a vocation. In H. H. Gerth & C. W. Mills (Eds.), *From Max Weber. Essays in sociology* (pp. 129–156). London: Routledge & Kegan Paul.

Index

B. Tholen, *Virtue and Responsibility in Policy Research and Advice*,
DOI 10.1007/978-3-319-65253-5

The manufacturer's authorised representative in the EU is Springer Nature Customer Service Centre GmbH, Europaplatz 3, 69115 Heidelberg, Germany. If you have any concerns regarding our products, please contact ProductSafety@springernature.com

Printed and bound by CPI Group (UK) Ltd, Croydon, CR0 4YY
30/09/2026
02225666-0002